D1315559

Associative N₂-Fixation

Volume II

Editors

Peter B. Vose, Ph.D.
Project Manager
Centre de Energia Nuclear
na Agricultura
Piracicaba, Brazil

Alaides P. Ruschel, Ph.D.
Centre de Energia Nuclear
na Agricultura
Piracicaba, Brazil

CRC Press, Inc.
Boca Raton, Florida

Library of Congress Cataloging in Publication Data

International Workshop on Associative N$_2$-Fixation,
    University of São Paulo, 1979.
    Associative N$_2$-fixation.

    Includes bibliographies and index.
    1. Nitrogen—Fixation—Congresses. 2. Micro-
organisms, Nitrogen-fixing—Congresses. I. Vose,
Peter B. II. Ruschel, Alaides P. III. Title.
S651.I55 1979   589.9'5'04133   80-23809
ISBN 0-8493-6130-3 (v. I)
ISBN 0-8493-6131-1 (v. II)

This book represents information obtained from authentic and highly regarded sources. Reprinted material is quoted with permission, and sources are indicated. A wide variety of references are listed. Every reasonable effort has been made to give reliable data and information, but the author and the publisher cannot assume responsibility for the validity of all materials or for the consequences of their use.

Direct all inquiries to CRC Press, Inc., 2000 N.W. 24th Street, Boca Raton, Florida 33431.

International Standard Book Number 0-8493-6130-3 (Volume I)
International Standard Book Number 0-8493-6131-1 (Volume II)

Library of Congress Card Number 80-23809
Printed in the United States

Proceedings of the International Workshop of Associative N₂-Fixation, July 2-6, 1979, held at Centro de Energia Nuclear na Agricultura, University of São Paulo, Piracicaba, Brazil, with the cooperation of Programa Fixação Biológica de Nitrogênio (CNPq), EMBRAPA, Serviço Nacional de Levantamento e Conservação de Solos, Km 47, Rio de Janeiro*

PROGRAM SESSIONS AND CHAIRMEN

Azospirillum and General Studies
Chairmen: A. P. Ruschel (morning)
            J. Döbereiner (afternoon)

Wheat, Maize and Sorghum
Chairmen: A. H. Merzari (morning)
            D. H. Hubbell (afternoon)

Grasses
Chairman: N. S. Subba Rao (morning)

Sugarcane
Chairman: D. G. Patriquin (afternoon)

Sugarcane (continued)
Chairman: M. Fried (morning)

Rice
Chairman: M. C. C. McDonald Dow (afternoon)

Evaluation Sessions
Chairman: P. B. Vose (morning)

* A number of papers presented at the Workshop could not be published here, because of prior submission to journals or for other reasons. They include:

> Nitrogen Balance in Flooded Rice
> A. App
> Carbon Metabolism of *Spartina alternifolia* Roots and of
> Root and Rhizosphere Diazotrophs.
> C. D. Boyle and D. G. Patriquin

INTRODUCTION

As its name implies, the International Workshop of Associative N_2-Fixation was a "working" meeting where there was a great deal of input from the participants, the majority of whom presented papers. Special features of the Workshop were the discussion sessions and the preparation of summaries of the progress in this research area.

Compared with research on nitrogen fixation by the legume-*Rhizobium* system, work on the associative systems is at a very early stage, despite the fact that Lipman and Taylor claimed evidence for N_2-fixation in connection with wheat as early as 1922. However, it is only within recent years that it has been confirmed that grasses, sugarcane, and certain wheat and maize varieties will support N_2-fixation. The nitrogen fixing water fern, *Azolla,* is now also attracting much research in connection with rice production. These various nitrogen fixing systems have been loosely termed "associative", although is is now clear that the different systems seem to have their own unique features.

The overall objective of most current research is to see if associative N_2-fixing systems can be developed to the point where they can make a worthwhile contribution to crop production or reduce the need for nitrogenous fertilizers. In the short term the benefits of associative nitrogen fixation are likely to be especially relevant to relatively unsophisticated low-cost farming systems, where cost of N-fertilizer is a major handicap. After the intensive efforts made by many workers in the last few years it now seemed an appropriate time to discuss the results, overall progress, method, and future research priorities. Despite the many meetings on nitrogen fixation in the last few years, this appears to have been the first one entirely devoted to associative N_2-fixation systems, or diazotrophic rhizocoenosis, to use the new terminology.

There have been some problems of terminology, and in order to resolve some of these difficulties there was an open discussion on the topic. Dr. W. J. Payne kindly acted as Chairman.

It was clear from the papers submitted that although there was some overlapping the work could be divided into four major area of interest, largely on a crop basis: *Azospirillum* and grass studies; wheat, maize and sorghum; rice; and sugarcane. Discussion of progress and priority research needs was therefore carried out initially on a "working party" basis prior to discussion before a final plenary meeting. The following agreed to act as Chairman of the four working parties:

Azospirillum and grass studies:	Dr. Johanna Döbereiner
Wheat, maize and sorghum:	Dr. R. V. Klucas
Rice:	Dr. I. Watanabe
Sugarcane:	Dr. Alaides P. Ruschel

Each group prepared a short paper outlining its conclusions as to the present position and the areas where research effort is especially needed. During the event it was decided that there was so much in common among the *Azospirillum* and grass studies, and the wheat, maize, and sorghum work that a single joint statement should be prepared.

For the final session, Dr. R. J. Rennie kindly agreed to put together, and present, a Workshop consensus paper which attempts to summarize the overall conclusions of the Workshop, review progress, and identify general areas that now seem to require most effort. The "position summaries" prepared by each working party and the consensus paper comprise the final section of Volume II. It is hoped that they may serve as a guide and stimulus to further work in this area.

In general the papers were published here both in the order and in the Section in which they were presented to the Workshop. However, in a number of cases papers have been moved to a more appropriate grouping.

As editors of the Proceedings we should like to thank everyone who helped to make the Workshop a success and especially Mrs. Diva Athié, Senior Administrative Assistant UNDP/IAEA Project BRA/78/006 for all her invaluable work, and also the Publisher, CRC Press, Especially Ms. Benita Budd, our Coordinating Editor.

Peter B. Vose

Alaides P. Ruschel

ACKNOWLEDGMENTS

It would not have been possible to hold the Workshop without the support of many organizations. The cooperation and assistance of the following organizations is greatly appreciated:

Comissão Nacional de Energia Nuclear (Brazil)
Secretaria de Cooperação Técnica Internacional — Subin (Brazil)
Conselho Nacional de Desenvolvimento Cientifico e Tecnológico — CNPq (Brazil)
United Nations Development Program — UNDP
International Atomic Energy Agency — IAEA
Fundação de Amparo à Pesquisa do Estado de São Paulo — FAPESP
National Sciences Foundation (USA)   Exchange
National Environmental Research Council (Canada)   Programs

The presentation of a copy of the Proceedings to each participant was made possible by the generous support of Fundação de Amparo à Pesquisa do Estado de São Paulo.

EDITORS

P. B. Vose was born in Manchester, England and attended the University of Glasgow. He has researched primarily on genetical effects in plant nutrition, isotope techniques *(Introduction to Nuclear Techniques in Agronomy and Plant Biology),* and latterly on nitrogen fixation. Although the author of about 60 papers he has spent much of his career in scientific administration and public affairs. He has worked in the United Kingdom, in California, in South Korea, in the secretariat of the International Atomic Energy Agency, Vienna, and is currently in Brazil for IAEA.

Alaides P. Ruschel was born in Araguaia, Brazil and attended the Agronomy School of the Federal Rural University of Brazil, Rio de Janeiro. She earned an M.S. at Purdue University, Indiana, and a Ph.D. at Escola Superior de Agricultura "Luiz de Queiroz" (ESALQ), Piracicaba, São Paulo, Brazil. Her main field of research is on biological nitrogen fixation and nitrogen nutrition of plants, major studies being nitrogen fixation in *Phaseolus,* and associated with sugarcane. She is at present Head of the Soil Microbiology Section of Centro de Energia Nuclear na Agricultura (CENA), Piracicaba, SP, Brazil.

TABLE OF CONTENTS

Volume I

SECTION I: *AZOSPIRILLUM* AND GENERAL STUDIES

SECTION II: WHEAT, MAIZE, AND SORGHUM

Volume II

SECTION I: GRASSES

SECTION II: SUGARCANE

SECTION III: RICE

SECTION IV: EVALUATION SESSIONS

Chapter 1

PROSPECTS FOR INOCULATION OF GRASSES WITH *AZOSPIRILLUM* SPP.

J. Döbereiner* and Vera Lucia D. Baldani**

TABLE OF CONTENTS

*   Programa Fixacao Biologica de Nitrogenio, Rio de Janeiro, Brazil.
** Graduate student of the Universidade Federal Rural do Rio de Janeiro.

I. INTRODUCTION

The title of this paper was proposed with the intention of stimulating discussion during this meeting which originally was thought to be a workshop with a small group of participants directly concerned with nitrogen fixation in rhizocoenosis. Interest however was so great that the scope of the meeting was widened. We still decided to maintain this paper although it contains few original data on its own and rather tries to interpret, from a different angle, data already presented in this meeting or elsewhere.

The urgency to obtain grasses and cereals which can cover part of their nitrogen needs by biological fixation precipitated attempts for premature solutions which had no scientific background. It is the tendency of the layman to attribute to the soil microbiologist the main and almost exclusive responsibility of inoculating plants with microorganisms. Even in the well established legume symbiosis, inoculation seldom produces drastic yield increase in the field. This is because the addition of relatively small numbers of selected superior *Rhizobium* strains has little effect if there is no establishment and selective multiplication of the inoculated bacteria in the rhizosphere. Even if the inoculated strains are established, they will only bring about yield increases if there are no similar bacteria already present in the soil and if environment and plant genotype permit maximal N_2-fixation. Date[4] found that tropical forage legumes normally do not produce more than 5% of nodules with the inoculated strain.

More than in legumes, the success of inoculation of grasses is dependent on the above mentioned conditions and identification of limiting factors, and attempts to find agronomically viable practices to eliminate them seem more promising research goals than simple inoculation trials. The recent advance in the understanding of the *Azospirillum* spp. associations are in support of this concept. Before one can think about producing inoculants it is essential to understand host plant specificity[2] and the plant and bacterial physiology[12,18,20] which interfere in root infection and principally N_2-fixation. The promotion of more stable and constant N_2 - fixing associations seems first priority. The discovery of denitrification by *Azospirillum lipoferum*[10] and of the role of *Azospirillum* spp. in nitrogen assimilation by plants[9,19] give additional alarming examples of possible hazards due to inoculation.

II. INOCULATION EXPERIMENTS

The announcement of N_2-fixation in a few maize lines in Brazil[21], which had not been inoculated but from which N_2-fixing *Azospirillum* was isolated, motivated a large number of inoculation experiments,[3,16,17] mainly in the U.S. The strain Sp 7 used in most of these experiments was the type strain (ATCC 29145) isolated by us from soil under *Digitaria* pastures. Statistical analyses of some of these experiments revealed significant increases in plant total N (around 15 kg N/ha) in some and significant decreases in others.[1] Figure 1 was prepared from data of four of these experiments carried out in Wisconsin. Highly significant ($p < 0.001$) negative correlations of the inoculation effect with the available soil N, as measured by N% in control plants, were obtained and it seems that in soils well provided with nitrogen, inoculation can lead to negative effects, possibly due to denitrification. It is now well known that many *Azospirillum* strains, Sp 7 among them, are very active denitrifiers.

These results represent one typical example of inoculation problems with cereals or grasses and if they ever should lead to success, extensive strain selection and optimization of environmental conditions are an absolute prerequisite. The various factors which lead to maximal N_2-fixation must be considered in step by step approaches, where the most perfect legume symbiosis can be used as a model system.

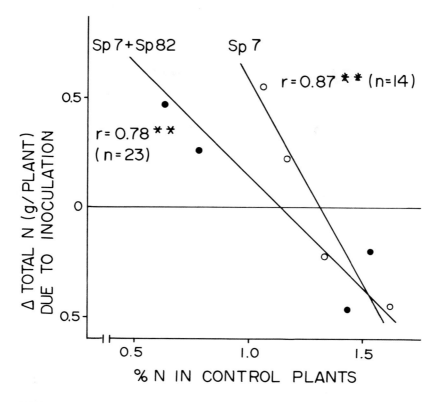

FIGURE 1.   Effects of inoculation of maize with *Azospirillum brasilense* on nitrogen incorporation under field conditions (recalculated from Albrecht[1]). The correlations were calculated from differences between inoculated and control plants using means of groups of genotypes (hybrid maize, inbred lines, field corn). Mean values of the four experiments are shown. The inoculation consisted of application of 3 mℓ/plant of NH$_4$ - grown *Azospirillum brasilense* (10^6 to 10^7 organisms cells/mℓ) at planting and again at thinning, and the controls were treated with heat killed inoculant.

The very first step is the establishment of the inoculated bacteria in the rhizosphere, and on and in the roots. Table 1 gives examples of inoculation effects on *Azospirillum* numbers. Under Wisconsin conditions, where no *Azospirillum* was present in the soil, inoculation seemed successful in terms of establishment. Under Brazilian conditions, increases in numbers of *Azospirillum* by inoculation of nonselected strains are less impressive, especially within roots. This would be expected due to the high numbers of spontaneous *Azospirillum* in the soil.

III. CHARACTERISTICS OF SUPERIOR STRAINS

In speaking of establishment of superior strains one has to define first which characteristics are really desirable. As in *Rhizobium,* selection of efficiency in nitrogen fixation in vitro has little relation with symbiotic performance. Pronounced host plant specificity in *Azospirillum* associations has been shown in several papers in this symposium,[2,5,14] and it seems that *A. lipoferum* infects C$_4$ plants and *A. brasilense* nir$^-$ C$_3$ plants. An example is given in Figure 2. Possibly subgroups within *A. brasilense* nir$^-$ further divide the *Azospirillum* spp.[5] Higher incidence of nir$^-$ *A. lipoferum* strains among isolates from sterilized maize roots when compared with soil[2] indicates subdi-

Table 1

EFFECT OF *AZOSPIRILLUM* INOCULATION
ON ESTABLISHMENT IN SOIL AND ROOTS.
DATA ARE NUMBERS OF *AZOSPIRILLUM*
SPP. × 10^5/pg ROOTS OR SOIL, PLANTS
BEING HARVESTED AT REPRODUCTIVE
GROWTH STAGE

Sample	Inoculated	Uninoculated check
Maize in Wisconsin[a]		
Surface ster. roots	0.001	0
Surface ster. crushed roots	65.0	0
Maize in Rio de Janeiro State[b]		
Soil	3.25	0.83
Roots	77.50	28.0
Surface ster. crushed roots	0.04	0.02
Sorghum in Rio de Janeiro[c]		
Roots	51.6	46.2
Surface ster. crushed roots	33.8	16.6
Wheat in Rio de Janeiro[d]		
Soil	1.07	0.78
Roots	19.33	6.70
Surface steril. crushed roots	3.34	0.34
Rice in Rio de Janeiro[d]		
Soil	1.97	0.45
Roots	4.70	5.09
Surface ster. crushed roots	0.03	0.02

[a] Means of two replicates from one maize root.[11] Strains isolated from soils were used as inoculants.
[b] Means of four plants collected in the field. Strains isolated from sterilized roots of homologous plants were used as inoculants. (From unpublished data of Baldani, V. L. D. and Döbereiner, J.)
[c] Means of five plants collected in the field[a] (Data from Freitas)
[d] Means of four plants collected in the greenhouse. Strains isolated from nonsterilized roots were used as inoculants. (From unpublished data of Baldani, V. L. D. and Döbereiner, J.)

vision in this group as well (Table 2). In addition maize and wheat root infection seems to occur preferentially by streptomycin resistant *Azospirillum* strains[6,7] (Figure 3).

Another important question in deciding which kind of *Azospirillum* strains are most suitable for inoculation is their ability to reduce NO_3^- to NO_2^-, or to dissimilate NO_3^- to N_2 gas. There are variations among *Azospirillum* strains and so far most strains isolated from surface sterilized roots are either *A. brasilense* nir⁻ or *A. lipoferum* nir⁻ (see also Table 2). Furthermore, mutants of *Azospirillum* which are nitrate reductase negative (nr⁻) are able to fix N_2 in the presence of NO_3^- in culture medium and in plants (Table 3). On the other hand, it has also been shown in several experiments that low doses of NO_3^- (5 to 10 kg/ha) increase nitrogenase activity in grass roots infected by *Azospirillum*[13] which could be attributed to NO_3^- dependent nitrogenase activity under oxygen limiting conditions.[15] The significant differences in leaf nitrate reductase found between plants inoculated with various *Azospirillum* strains in rice,[19] and in sorghum,[9] give additional support for the role of *Azospirillum* in nitrate metabolism of the plant.

Before all these interactions of *Azospirillum* with plants are well understood it seems

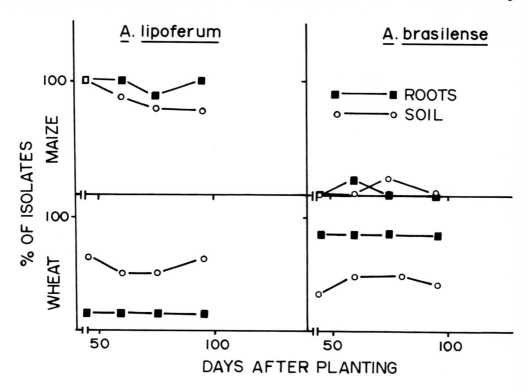

FIGURE 2. Host plant specificity in the infection of maize and wheat roots with *Azospirillum* spp. *Azospirillum* was isolated from surface sterilized roots (1% chloramine-T for 1 hr for maize and 15 min for wheat) and soil, at four growth stages of field grown plants, and identified according to Tarrand.[22] The percentage of each species was calculated from two isolates each from four field plots. (Data reformulated from Baldani, V. L. D. and Döbereiner, J.[2])

Table 2
SPONTANEOUS SELECTION FOR NONDENITRIFYING (NIR⁻) *AZOSPIRILLUM LIPOFERUM* IN SURFACE STERILIZED MAIZE ROOTS[a]

	Washed roots	Roots ster. 30 sec.[b]	Roots ster. 60 min[b]
		% nir⁻ strains	
Greenhouse			
Flowering	33	3	50
Grain filling	0	50	50
Maturation	0	33	3
Field experiment			
Flowering	0	10	63
Grain filling	10	3	50
Maturation	3	3	10

[b] One percent chloramine T.

[a] According to Tarrand[22] 92% of the *Azospirillum lipoferum* strains are denitrifying (nir⁺)

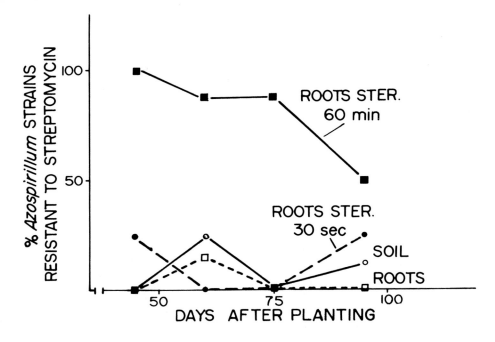

FIGURE 3.   Spontaneous occurrence of streptomycin resistant (20 μg/mℓ) *Azospirillum lipo-ferum* strains in field grown maize roots. *Azospirillum* was isolated from rhizosphere soil and from not sterilized, 30 sec sterilized, and 60 min sterilized maize roots and *Azospirillum* spp. and streptomycin resistance determined according to Döbereiner and Baldani,[6] The percentage was calculated from two isolates each from four field plots.

Table 3

NITROGENASE ACTIVITY (C_2H_2) OF NITRATE (NR^-) AND
NITRATE (NIR^-) REDUCTASE NEGATIVE MUTANTS OF
AZOSPIRILLUM BRASILENSE IN THE PRESENCE OF NO_3^-

	n mol C_2H_4/hr/culture		
	Culture medium[a]		Monoxenic plants[b]
	(without NO_3^-)	(with NO_3^-)	(with NO_3^-)
Parent strain nr⁺ nir⁺	120	1	38
Mutant nr⁺ nir⁻	180	2	33
Mutant nr⁻ nir⁺	260	110	69
Mutant nr⁻ nir⁻	270	246	89

[a]   Semisolid malate cultures with or without 10 mMKNO₃.
[b]   Sorghum seedlings cultivated in sterile vermiculite with nutrient solution containing 1 mMKNO₃ and inoculated with the various *Azospirillum* mutants. Nitrogenase activity measured in intact systems under pO_2 = 0.005 atm.

From Döbereiner, J., *Interciencia*, 1979b, in press.

rather a shot in the dark to inoculate field plants with *Azospirillum strains*. In several experiments in which we tried to use strains isolated from homologous roots which were selected for streptomycin resistance, variable results were obtained. Maize,[2] sorghum,[9] and rice[19] when inoculated with such a strains showed no plant N increases and in some cases even decreases.

Table 4
INCIDENCE OF *AZOSPIRILLUM* SPP. IN WHEAT AND MAIZE ROOTS AS EVALUATED BY NITROGENASE ACTIVITY IN LOG PHASE ENRICHMENT CULTURES. DATA ARE FROM A FIELD EXPERIMENT WITH HEAVY *AZOSPIRILLUM* INOCULATION. MEANS OF FOUR HARVESTS WITH FOUR REPLICATES EACH IN NMOL C_2H_4/HR/CULTURE (4 Ml)

Sample	Sterilized in chloramine-T (min)	Uninoculated check	Inoc. with maize strain[a]	Inoc. with wheat strain[b]
Wheat				
Soil	0	71	169	166
Roots	0	147	1038	817
Roots	0.25	66	214	171
Roots	15	54	88	400
L.s.d. (p = 0.05) for the interaction is 96				
Maize				
Soil	0	206	158	109
Roots	0	66	79	37
Roots	0.5	16	8	11
Roots	60	4	4	12
L.s.d. not significant				

[a]   *A. lipoferum* st^r isolated from 60 min sterilized maize roots.
[b]   *A. brasilense* nir$^-$ st^r isolated from 15 min sterilized wheat roots.

In one experiment with wheat, however, a promising host strain combination seems to have been found. In this experiment the inoculated homologous strain established within the roots (Table 4). Another strain which had been isolated from maize established equally well in soil and on the root surface but not within the roots. In contrast, there were no inoculation effects on the incidence of *Azospirillum* in maize roots whether the strains were from wheat or maize.

Accordingly, only in the wheat experiment were inoculation effects observed on plant growth (Table 5). Plants inoculated with the streptomycin resistant wheat strain showed not only higher straw and grain yields but especially increased total N in plants. These increases correspond to increases of 16, 30, and 35%, respectively, of plant dry weight, total plant nitrogen, and grain yield.

Whether the increase in plant N is due to N_2-fixation cannot be said, because it has been shown in this symposium that *Azospirillum* can influence nitrogen incorporation by plants also by other means (see above).

IV. CONCLUSIONS

In conclusion it can be said that recent results have increased the chances of plant benefits from *Azospirillum* inoculation. Although simple inoculation of any strain will most probably give no or negative results, a large number of research lines have opened which promise to lead eventually to success, at least in tropical or subtropical regions. Among these research objectives the following are stressed as most promising:

1.   Better understanding of host plant specificities
2.   Identification of groups by immunofluorescence antibody techniques

Table 5
EFFECT OF INOCULATION OF FIELD GROWN WHEAT
WITH *AZOSPIRILLUM* SPP. ISOLATED FROM SURFACE
STERILIZED MAIZE OR WHEAT ROOTS. DATA ARE
MEANS OF FOUR HARVESTS AND FOUR REPLICATES
EACH OF TWO PLANTS PER PLOT. GRAIN YIELDS ARE
MEANS OF TWO 1 M ROWS FROM EACH OF FOUR PLOTS

Inoculant	Dry wt /two plants(g)	N %	Total N in two plants (mg)	Grain yield g/l m row
A. lipoferum st^r from maize	5.29	1.89	96.1	11.21
A. brasilense nir$^-$ st^r from wheat	6.34	1.90	117.9	11.75
Check	5.46	1.73	91.0	8.73
L.s.d. ($p = 0.01$)	0.25	n.s.	6.3	n.s.

3. Determination of levels and forms of antibiotics within roots and the require-
    ments for resistance of infecting bacteria
4. Manipulation of the rhizosphere population by the introduction of certain acti-
    nomycetes and *Azospirillum* strains resistant to the antibiotics produced by them
5. Use of nr$^-$ and nir$^-$ mutants to establish the role of *Azospirillum* in nitrate metab-
    olism of the plant
6. Establishment of systems which can use mineral nitrogen and N₂ simultaneously

REFERENCES

1. **Albrecht, S. L., Okon, Y., Lonquist, J., Döbereiner, J., and Burris, R. H.,** unpublished data; cited in **Döbereiner, J.,** Nitrogen fixation in grass-bacteria associations in the tropics, *Isotopes in Biological Dinitrogen Fixation,* International Atomic Energy Agency, Vienna, 1978, 51.
2. **Baldani, V. L. D. and Döbereiner, J.,** Host plant specificity in the infection of maize, wheat and rice with *Azospirillum* spp., in *Associative N₂-Fixation,* Vose, Peter B. and Ruschel, A. P., Eds., CRC Press, Boca Raton, Fla., 1981.
3. **Burris, R. H.,** Potential of associated nitrogen fixing systems, in *Genetic Engineering for Nitrogen Fixation,* Hollaender, A. Ed., Plenum Press, New York, 1977, 443.
4. **Date, R. A.,** Nitrogen, a Major Limitation in the Productivity of Natural Communities, Crops and Pastures in the Pacific Area, 12th Pacific Sci. Cong., Canberra, Australia, 1971.
5. **de-Polli, H., Bohlool, B. B., and Döbereiner, J.,** Immunofluorescence Differentiation of *Azospirillum* Species Belonging to Different Host-Plant Specificity Groups, Int. Workshop Associative N₂-Fixation, Piracicaba, Brasil, 1979.
6. **Döbereiner, J. and Baldani, V. L. D.,** Selective infection of maize roots by streptomycin resistant
7. **Döbereiner, J. and Baldani, V. L. D.,** Increase of Antibiotic Resistant Bacteria in Grasses Roots, Int. Workshop Associative N₂-Fixation, Piracicaba, Brasil, 1979.
8. **Döbereiner, J.,** Fixação de nitrogênio em gramíneas tropicais, *Interciência,* 1979b, in press.
9. **Freitas, J. L. M., Pereira, P. A. A., and Dobereiner, J.,** Effect of Organic Matter and *Azospirillum* Strain on N Metabolism in *Sorghum vulgare,* Int. Workshop Associative N₂-Fixation, Piracicaba, Brasil, 1979.
10. **Neyra, C. A., Döbereiner, J., Lalande, R., and Knowles, R.,** Denitrification by N₂-fixing *Spirillum lipoferum, Can. J. Microbiol.,* 23, 300, 1977.
11. **Okon, Y., Albrecht, S. L., and Burris, R. H.,** Methods for growing *Spirillum lipoferum* and for counting it in pure culture and in association with plants, *Appl. Environ. Microbiol.,* 33, 85, 1977.

12. Pedrosa, F. O., Stephan, M. P., and Döbereiner, J., Interaction of Nitrogenase Activity and uptake and Hydrogenase in *Azospirillum brasilense*, Int. Workshop Associative N₂-Fixation, Piracicaba, Brasil, 1979.

13. Pereira, P. A. A., Neyra, C. A., and Döbereiner, J., Atividade da nitrogenase, nitrato redutase e níveis de nitrato em 5 genótipos de *Brachiaria* spp., 17th Congr. Brasileiro de Ciência do Solo, Manaus, Brasil, 1979.

14. Rocha, R. E. M., Baldani, J. I., and Döbereiner, J., Host Plant Specificity in the Infection of C₄ Plants by *Azospirillum* spp. Int. Workshop Associative N₂-Fixation, Piracicaba, Brasil, 1979.

15. Scott, D. B., Scott, C. A., and Döbereiner, J., Nitrogenase activity and nitrate respiration in *Azospirillum* spp., *Arch. Microbiol.*, 121, 141, 1979.

16. Sloger, C. and Owens, L. D., N₂-Fixation by a Temperate Corn-*Spirillum* sp. Association, 2nd Int. Symp. N₂-Fixation, Salamanca, Spain, 1976.

17. Smith, R. L., Bowton, J. H., Schank, S. C., Quesenberry, K. H., Tyler, M. E., Milam, J. R., Gaskins, M. H., and Littele, R. C., Nitrogen fixation in grasses inoculated with *Spirillum lipoferum*, *Science*, 193, 1003, 1976.

18. Stephan, M. P., Pedrosa, F. O., and Döbereiner, J., Physiology Studies of *Azospirillum* spp., Int. Workshop Associative N₂-Fixation, Piracicaba, Brasil, 1979.

19. Villas Boas, F. C. S. and Döbereiner, J., Efeito de Diferentes Níveis de N Mineral na Atividade da Nitrato Redutase e Nitrogenase em Arroz. (*Oriza sativa*) Inoculado com duas Estirpes de *Azospirillum* spp., Int. Workshop Associative N₂-Fixation, Piracicaba, Brasil, 1979.

20. Volpon, A. G. T., de-Polli, H., and Döbereiner, J., Changes in Efficiency of Nitrogen Fixation in Various Growth Stages of Batch Cultures of *Azospirillum lipoferum*, Int. Workshop Associative N₂-Fixation, Piracicaba, Brasil, 1979.

21. von Bülow, J. F. W. and Döbereiner, J., Potential for nitrogen fixation in maize genotypes in Brasil, *Proc. Nat. Acad. Sci. U.S.A.*, 72, 2389, 1975.

22. Tarrand, J. J., Krieg, N. R., and Döbereiner, J., Taxonomic study of *Spirillum lipoferum* group, with description of a new genus, *Azospirillum* gen nov. and two species *Azospirillum lipoferum* (Beijerinck) comb. nov. and *Azospirillum brasilense* sp. nov., *Can. J. Microbiol.*, 24, 967, 1978.

Chapter 2

PHYSIOLOGY OF THE DIAZOTROPHIC RHIZOCOENOSIS IN SALT MARSH CORD GRASS, *SPARTINA ALTERNIFLORA* LOISEL

D. G. Patriquin, C. D. Boyle, D. C. Livingstone, and C. R. McLung*

TABLE OF CONTENTS

*   Biology Department, Dalhousie University, Halifax, N.S., Canada.

I. INTRODUCTION

Observation of a large net export of organic N from a small coastal inlet in Nova Scotia[1] led to a search for possible sites of N_2-fixation, and recognition of an "associative symbiosis" in *Spartina alterniflora*.[2] This C-4 type, perennial grass is the primary subaerial colonizer of intertidal mudflats in eastern North America.

This presentation is a review of our work concerned with physiological interactions between *Spartina* and associated bacteria. One object was to determine which of the several diazotrophic bacteria associated with the roots is responsible for nitrogenase activity. Döbereiner and Day[3] proposed that nitrogenase activity in certain tropical grasses is associated with internal infections by *Spirillum lipoferum* (syn. *Azospirillum* spp.[4]). This diazotroph is microaerophilic under N_2-fixing conditions, and grows on common organic acids as sole carbon sources. Döbereiner and Day[3] suggested that the latter characteristic may facilitate a close association with C-4 grasses. Our observations suggest that internally located, organic acid-utilizing microaerophils are also responsible for root nitrogenase activity in *Spartina*. However, the root infections are mixed ones in which the microaerophils are a relatively minor component numerically, and thus the plant-bacterial carbon relationships are more complex than envisaged.[3]

II. BACTERIA

Nitrogenase activity by roots excised from carbon starved *Spartina* plants (plants kept in the dark) is stimulated by both malate and glucose.[5] Several facultatively anaerobic, glucose-utilizing diazotrophs, and a malate-utilizing microaerophilic diazotroph were isolated from the roots.[2] The facultative organisms, which do not grow on malate as a sole carbon source, have not been identified. Our first attempts to maintain a microaerophil isolated from roots of *Spartina*[2] were unsuccessful, the organism being lost after several transfers on plates. More recently we have isolated a salt-requiring, obligately microaerophilic diazotroph from a 10^5 dilution of chloramine-T treated roots, and have been able to maintain this organism in semisolid malate medium at room temperature by transferring it to fresh medium at approximately weekly intervals. The low G + C ratio of this organism (32.1 ± 1), its obligately microaerophilic habit, and lack of PHB (poly-B-hydroxybutyrate) indicate that it is a species of the genus *Campylobacter*.[33] Sources of presently described members of this genus are mammalian tissues of various sorts.[6] The isolate from *Spartina* has a single polar flagellum, is motile, and grows on common organic acids (aspartate, fumarate, lactate, malate, pyruvate, succinate) except for citrate, as sole carbon sources. It does not utilize common sugars (sucrose, fructose, glucose) as sole carbon sources. This appears to be the first report of a microaerophilic diazotroph other than *Azospirillum* spp. being isolated from grass roots.

Immersion of roots of *Spartina* in 1% chloramine-T for 2 hr kills all cells in the epidermis and outermost three to four layers of the cortex.[5] Thus the bacterial population recovered after chloramine-T treatment is an internal population. The proportion of microaerophils in this internal population is 16 to 19 times greater than their proportion in the whole root population (Table 1), showing that there is a "relative enrichment" of microaerophils in the interior of the root, or "endorhizosphere". There is also a relative enrichment of the facultative anaerobes, but by a much smaller factor. The diazotrophs collectively make up less than 10% of the endorhizosphere population (Table 1). Similarly, diazotrophs constitute a small proportion of the populations recovered after surface sterilization of wheat roots,[7] roots of maize inoculated with *Azospirillum*,[3] and stems of sugar cane.[9]

Table 1
PROPORTION OF DIAZOTROPHS IN BACTERIAL POPULATION OF SOIL, WASHED ROOTS, AND SURFACE STERILIZED ROOTS[5,16]

Sample	Total No./g wet weight (plate count)	% diazotrophs		Relative enrichment of diazotrophs in root interior	
		Glucose-utilizing anaerobes	Malate-utilizing microaerophils	Anaerobes	Microaerophils
Field samples					
Nonrhizosphere soil	1.09×10^6	0.012	0.022		
Rhizosphere soil	28.8×10^6	3.2	0.32		
Washed roots	34.7×10^6	2.6	0.081		
Surface sterilized roots[a]	0.10×10^6	5.4	1.3	2.0	16
Plants in hydroponics[b]					
Washed roots	16.1×10^6	2.9	0.022		
Surface sterilized roots[a]	0.69×10^6	6.7	0.41	2.3	19

[a] One percent chloramine-T × 2 hr.

[b] Field plants were transferred to flowthrough seawater-hydroponics; counts after approximately 6 months.

III. INFECTION

Bacteria in root tissues are not readily discriminated by light microscopy without staining, but are easily discriminated if they produce large formazan crystals in the presence of 2,3,5-triphenyltetrazolium chloride.[10] Bacteria producing such crystals include diazotrophs,[3] but not exclusively so.[10] Thus sites of tetrazolium-reducing bacteria in grass roots infected with mixed diazotroph/nondiazotroph populations represent possible sites of the diazotrophs. Clusters of cells in the outer cortex filled with tetrazolium-reducing bacteria (Plate 1) are conspicuous features of some *Spartina* roots.[2] Because the surrounding soils are generally anaerobic, and oxygen reaches the roots via diffusion through air spaces in the mid-cortical region (Plate 1), this outer cortex layer may be the zone of lowest pO_2 within the roots, i.e., be a locale favorable for microaerophils. Tetrazolium-reducing bacteria are also observed in other sites, including the surfaces of cells lining the air spaces, within the xylem and between pith cells in the stele, and aligned longitudinally along the inner cortex-outer stele region as in maize.[10] Patriquin and Döbereiner[10] cited the latter feature as evidence that lateral roots are sites of entry of bacteria into the roots. Such a mode of infection would explain our observation that nitrogenase activity of excised roots is more highly correlated with root age for unbranched roots than for branched roots (Figure 1), as the unbranched roots would have fewer loci for infection.

As in maize[10,11] and sugar cane,[9] these general internal infections in *Spartina* are established without radial disruption of outerlying tissues, or disruption of the endodermin. Although such observations refer to mixed populations, and not specifically to diazotrophs, studies with monoxenic grass-*Azospirillum* systems suggest that diazotrophs infect the same general sites.[10,12] It seems particularly significant that bacteria are commonly observed inside of the stele, as this is a region of low pO_2,[13] and occurrence of diazotrophs in this region would facilitate exchange of materials between the plant and bacteria. Initial entry into the stele probably occurs at root apices,[11] of which there are more in branched roots than in unbranched roots. A corollary of our interpretation of Figure 1 is that excised root nitrogenase activity is associated predominantly with internally located bacteria.

IV. NITROGENASE ACTIVITY OF ENDORHIZOSPHERE BACTERIA

The relative insensitivity of acetylene-reducing activity of roots excised from *Spartina* to differences in incubation pO_2 except under vigorous shaking (Figure 2), and the high apparent K_m for acetylene reduction by excised roots except after prolonged exposure to C_2H_2[14] also suggest that there must be a barrier of some sort to diffusion of gases (O_2 or C_2H_2) to nitrogenase sites, i.e., that the diazotrophs responsible for excised root acetylene-reducing activity are not simply sitting on the root surface. The applied significance of excised root nitrogenase activity is uncertain, however, because of the unexplained "lag" that precedes it.[14]

When leaves of intact plants in the field were enclosed in tubes above the soil and incubated with C_2H_2, production of C_2H_4 proceeded following a very short lag phase (Figure 3). No C_2H_4 production was observed when similar portions of leaves were excised and enclosed in tubes.[15] This suggested to us that in the intact system, C_2H_2 diffused down air spaces, and was reduced to C_2H_4 by bacteria in the vicinity of air spaces in the roots. Since C_2H_4 is relatively insoluble in water, ethylene produced would tend to diffuse back up the lacunae rather than to follow a path of greater resistance through the waterlogged soil. We postulated that if plants were enclosed in a two-phase, leaves-roots system (Figure 3) with C_2H_2, C_2H_4 produced by internally located

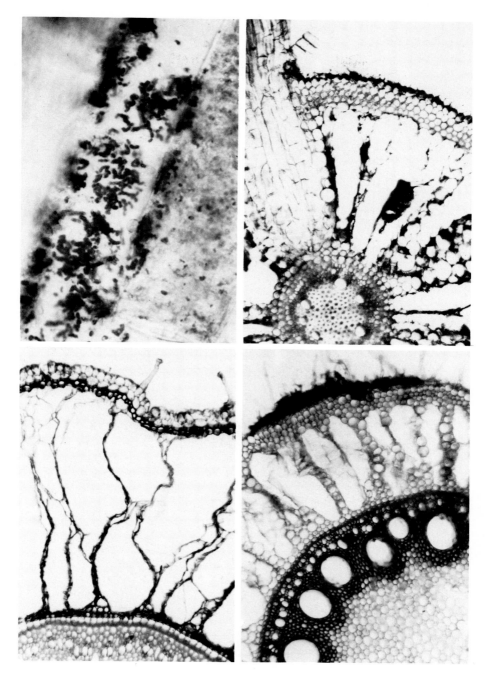

PLATE 1

Top left: Surface view of outer cortex of *Spartina* root illustrating bacteria with prominent formazan crystals around perimeter of cell. See Reference 2 for technique. (Magnification approximately × 1200.)

Top right: Cross section of *Spartina* root illustrating large air spaces. (Magnification approximately × 190.)

Bottom left: Cross section of root of sugar cane root illustrating large air spaces. Note heavily suberized outer cortex, or hypodermal layer. (Magnification approximately × 90.)

Bottom right: Cross section of maize root illustrating air spaces. (Magnification approximately × 90.)

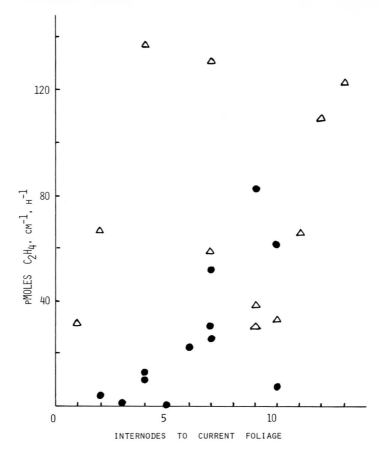

FIGURE 1. Acetylene-reducing activity (ARA) of excised roots in relation to relative age of roots (indicated by the number of nodes between node of root insertion foliage leaf node) for branched roots (triangles) and unbranched roots (circles) from a 1 + year old *Spartina* clump. (Data from Patriquin, D. G. and McClung, C. R., *Mar. Biol.*, 47, 227, 1978. With permission.)

bacteria would be detected in the upper (leaf) phase, while that produced by bacteria near the exterior or on the surface of the roots would be detected in the lower (root) phase, allowing us to differentiate between internal and external populations. Such a system would also allow ready maturation of both internal and external populations with C_2H_2.

Plants from the field were transferred to a flow-through seawater hydroponics system where they were maintained for several months before experiments of this nature were begun. Ethylene production was evident in both lower and upper phases within 1 to 2 hr of transferring plants to the two-phase assay systems (Figure 3). Addition of mercuric chloride to the lower phase resulted in an immediate 75% reduction in the lower phase C_2H_4 production, but had no immediate effect on upper phase C_2H_4 production. Acetylene-reducing activity of excised leaves amounted to 8 to 24% of the C_2H_4 production observed in the upper phase prior to excision of leaves.[16] Thus the upper phase C_2H_4 production seems to be attributable principally to nitrogenase activity of diazotrophs located inside of stems and roots. In various experiments, this upper phase or endorhizosphere acetylene-reducing activity, corrected for excised leaf activity

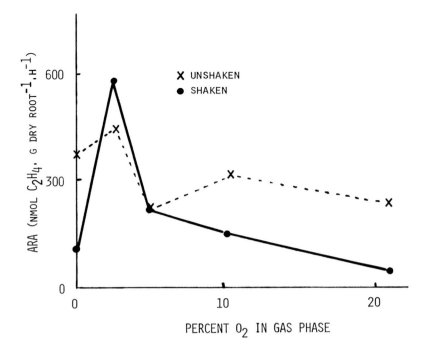

FIGURE 2. Effect of different oxygen concentrations, with and without shaking, on excised root acetylene-reducing activity. Roots were preincubated under 10% oxygen, and evacuated and backfilled with air and N_2 to give desired oxygen concentrations. Acetylene was injected, and C_2H_4 determined after 2 hr (unshaken). The roots, in flasks, were then put on a wrist action shaker which was set at high speed, and acetylene-reducing activity determined for the period 1.8 to 3.5 hr after shaking was initiated. Data points are means for duplicate samples. (Data from Patriquin, D. G., *Aquat. Bot.*, 4, 193, 1978. With permission.)

and for diffusion of C_2H_4 from the lower phase seawater, amounted to 10 to 85% of the whole plant acetylene-reducing activity, with the highest percentages being characteristic of the most active plants.[16]

Lower phase C_2H_4 production (in the presence of C_2H_2) was more responsible than that of the upper phase to addition or ammonium of mixed carbon sources to the lower phase seawater (Table 2). Similarly, for roots and rhizosphere soil taken from plants growing in pots, rhizosphere soil acetylene-reducing activity was more responsive than was excised root acetylene-reducing activity to additions of mercuric chloride, ammonium chloride, or mixed carbon sources. Infiltration of roots with the ammonium-containing seawater was facilitated in the excised root systems by evacuation. The lack of response of excised roots and upper phase (endorhizosphere) C_2H_4 production to addition of 200 μM ammonium is suggestive of some degree of coupling of plant bacterial metabolisms as in lower plant symbioses.[17]

V. EFFECTS OF OXYGEN

Nitrogenase activity of excised roots of *Spartina* may be described as O_2-sensitive/ O_2-requiring. Nitrogenase activity ceases or declines sharply soon after oxygen is depleted (Figure 3). Under conditions of vigorous shaking, nitrogenase activity is maximal at low pO_2 (Figure 2). The simplest interpretation is that microaerophils are responsible for most of the root nitrogenase activity. The persistence of nitrogenase

internal population of microaerophilic diazotrophs dependent on malate or other organic acids being responsible for root nitrogenase activity. Glucose-utilizing, facultative anaerobes may be active in N_2 fixation in the anaerobic soils surrounding *Spartina* roots.[14] The internal or endorhizosphere population is a mixed one, including a predominance of nondiazotrophic bacteria. Thus the plant-microbial carbon relationships could be expected to be more complex than envisaged by Döbereiner and Day.[3]

Boyle[16] found that acetylene-reducing activity of excised roots and rhizosphere soil taken from carbon starved *Spartina* plants (plants kept in the dark for 30 days) was stimulated by addition of macerates of roots from photosynthesizing plants, but not by such macerates when the ethanol soluble compounds were removed.

The principal constituents in the neutral ion exchange fraction of 80% ethanol extracts of *Spartina* roots are glucose, fructose, and sucrose, and in the acidic fraction are malate and citrate.[16,22]

Both malate and glucose stimulated acetylene-reducing activity of roots excised from carbon starved *Spartina* plants, but the glucose did so only after a long incubation period.[16] Glucose, but not malate, stimulated acetylene-reducing activity of rhizosphere soil.

Pulse exposure of photosynthesizing plants to 14-CO_2, and determination of the partitioning of 14-C after various chase periods showed that sugars are the predominant form in which recent photosynthates (i.e., synthesized within the previous 24 hr) are transported to roots and rhizosphere soil.[16] Metabolism of 14-C labeled neutral compounds in both excised roots and rhizosphere soils, after they were separated from labeled plants, resulted in a slow decline of label in the neutral fraction, with concommitant increase in labeling of the acidic and basic fractions.

Various experimental results indicate that recent photosynthates enter a large carbon pool within the root, that carbon substrates utilized by diazotrophs are derived from a large carbon pool, and that nitrogenase activity is proportional to the size of that root carbon pool:

1.   When leaves were pulse labeled with 14-CO_2 and roots removed after a 5 hr chase period, the specific activity of CO_2 respired over the ensuing 48 hr remained constant. The total 14-CO_2 evolved by 48 hr was equal to 25% of the 14-C remaining in the ethanol soluble fraction of the roots.[16]
2.   Nitrogenase activity of excised roots incubated under air,[14] and of intact plants in the dark[16] exhibited little decline over 10 days. After 29 days in the dark, intact plant nitrogenase activity (assayed intermittently) was still 50% as high as in plants subjected to normal photoperiods.
3.   For individual roots excised from a plant exposed to 14-CO_2, nitrogenase activity was proportional to root 14-C activity (Figure 4). Older roots had higher nitrogenase activity per unit 14-C activity than did younger roots, which is consistent with the concept that older roots would be more heavily infected by the appropriate diazotrophs.
4.   Seasonally, excised root acetylene-reducing activity on any one date was highly correlated with the root sugar concentration 2 weeks prior to that date (Figure 5).

Total sugar concentration in roots was as high as 29% of the dry weight,[22] while organic acids never accounted for more than 2 to 3% of the dry weight. Root nitrogenase activity was not correlated with the total organic acid content, but as *Campylobacter* does not utilize citrate, it is possible that root nitrogenase activity was proportional to root malate concentration. In any case, the malate would have to be derived from

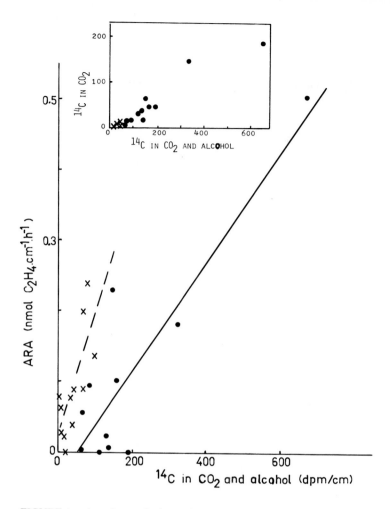

FIGURE 4. Acetylene-reducing activity (ARA) and radiocarbon activity of roots taken from a plant previously exposed to $^{14}CO_2$. *Spartina* clump was exposed to $^{14}CO_2$ for 4 hr at mid-day, dug up after 24 hr, and the individual roots incubated in syringes with C_2H_2 (5%), O_2 (5%) and N_2 (90%) for 24 hr; gas phase was then sampled for $^{14}CO_2$ and C_2H_4, and roots were extracted twice with boiling 80% ethanol. Inset shows that approx. 1/4 of the root ^{14}C activity had been respired after 24 hr. (From Patriquin, D. G. and McClung, C. R., *Mar. Biol.*, 47, 227, 1978. With permission.)

root sugars, and thus the correlation of root nitrogenase activity with sugar, rather than organic acid concentration, is not inconsistent with the idea that microaerophils are the major agents of root nitrogenase activity.

At low pO_2, organic acids may be made available from plant sugars through the fermentative action of facultative anaerobes, or through anaerobic metabolism of the plant itself (Figure 6). The latter scheme, in which malate is an end product of plant metabolism under O_2-deficient conditions[23] would require high rates of CO_2 fixation in the roots. For excised roots under 5% O_2, CO_2 fixation was estimated as between 0.639 and 1.78 μmol CO_2 fixed per gram fresh root per hour.[16] Assuming that this CO_2 is utilized in malate production and that the efficiency of N_2-fixation is equivalent to that of *Azospirillum* in pure culture,[24] the lower value would support acetylene-

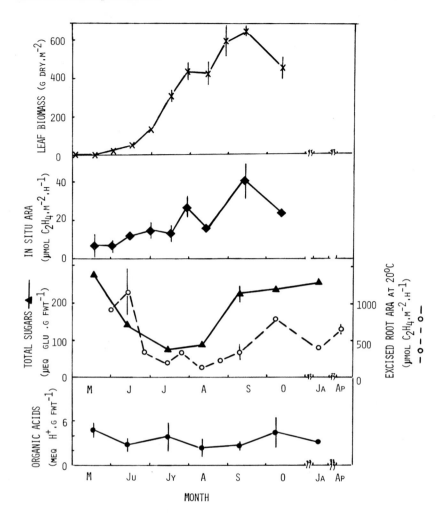

FIGURE 5.   Growth of *Spartina,* acetylene-reducing activity, and root total sugar and total organic acid concentrations at a thatch island *Spartina* stand in 1977. *In situ* ARA (acetylene-reducing activity) was measured using plexiglass cylinders;[15] algae and sand surface were scraped off and replaced with algae-free sand. Excised root samples consisted of 10 × 10 × 10 cm sods, assayed as in Reference 5. Only roots and stems attached to living foliage were analyzed for sugars and acids. (From Livingstone, D. C. Growth and Nitrogen Accumulation in *Spartina alterniflora* Loisel, in Relation to Certain Environmental and Physiological Factors, Master's thesis, Dalhousie University, Halifax, N.S., Canada, 1978. With permission.)

reducing activity of approximately 250 nmol C_2H_4/g wet root per hour if we convert N_2-fixation to acetylene reduction using a 1:4 *M* ratio. This value is well above most values observed for *Spartina* roots. Removal of CO_2 in homogenized excised root systems resulted in a four- to sixfold reduction in acetylene-reducing activity in comparison to controls under 1% CO_2,[16] suggesting that CO_2 is essential for root nitrogenase activity. The CO_2 requirement could, however, be due to some factor other than carbon source availability, for example pH regulation[25] or provision of carbon skeletons as acceptor molecules for fixed N_2[26]

Localized O_2 consumption by bacteria around root cells may induce a localized,

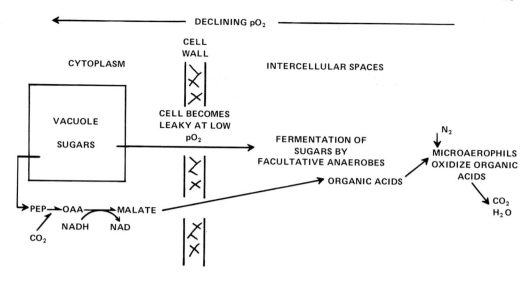

FIGURE 6. Possible means by which organic acids are produced from sugars in roots and utilized by microaerophilic diazotrophs.

rather than a general, leakage of sugars from plant cells. In order for microaerophils, which make up only about 1% of the endorhizosphere population, to metabolize a significant fraction of the organic acids produced in the root, it would seem to be necessary that they should be located in a region of slightly higher pO_2 than other aerobes (including the plant cells), or that they are able to utilize carbon compounds more effectively at low pO_2 than competing organisms. If the organic acids were metabolized in one of these modes, then most of the energy released during breakdown of root sugars, would be released during oxidation by the microaerophils. This could result in fairly high efficiencies of N_2-fixation. On the other hand, once the pO_2 increased sufficiently such that it was more readily utilized as a hydrogen acceptor by other cells, the efficiency of N_2-fixation would drop sharply. Sharply declining efficiencies of N_2 fixation or nitrogenase activity with increasing pO_2 might account for part of the discrepancy between acetylene-reducing activities measured *in situ* and those measured in excised root systems under low pO_2. This would account for the fact that these discrepancies are greater for "tall" *Spartina* stands,[22] than for "short" stands,[15] as the former are better aerated than the latter.

VII. SEASONAL PATTERNS OF NITROGENASE ACTIVITY

Our own,[5,22] and other[27] studies have consistently demonstrated a trend of increasing excised root acetylene-reducing activity through the growing season, with a maximum being reached in the fall during seed-fill. As indicated above, the excised root acetylene-reducing activity is proportional to, but lags slightly behind root sugar concentration. The changes in root sugar concentrations were typical of those in temperate zone perennial grasses,[28] with declining concentrations during early vegetative growth, and increasing concentrations in the fall. Neyra and Dobereiner[29] commented on similar seasonal trends in acetylene-reducing activity in other grasses, including maize, rice and sorghum, all of which exhibit a maximum at some time during reproductive growth. For those annuals, early season acetylene-reducing activity is lower, relative to the maximum, than in *Spartina,* and activity drops more rapidly following the maximum. These differences from *Spartina* are probably related to (1) the requirement for internal infections to first become established in the annuals,[11] and (2) rapidly declining root sugar concentrations during the grain-filling period in the annuals.[30]

The *in situ* acetylene-reducing activity exhibited a similar trend of increase through the season from June to the fall (Figure 5), but there were not high levels *in situ* in the spring as there were in the excised root systems. The *in situ* nitrogenase activity may be limited by absence of O_2 before growth begins (excised roots were incubated under 5% O_2), while high pO_2 may be the reason for the lower rates *in situ* in comparison to excised roots later in the season.

VIII. SOME IMPLICATIONS FOR THE ENHANCEMENT OF NITROGENASE ACTIVITY IN DOMESTICATED GRASSES

The question of whether or not internal infections of grasses are a significant factor in nitrogenase activity of grasses is a critical one, as the answer affects the way in which we go about seeking enhanced N_2-fixation. If the bacteria are principally external, then we might look for "leaky" plants, a lot of the N fixed could be expected to go into the soil initially, and the overall efficiency could not be very high. If the internal infections are important in grass N_2-fixation, then we are faced with quite different choices. For example our studies suggest that rather than being associated with leaky plants, high root nitrogenase activity will be associated with plants that accumulate a lot of sugars in the roots.

The relationships between root structure, root aeration and metabolism, and types and sites of infection may constitute a profitable area of investigation. Large air spaces are commonly associated with aquatic plants such as *Spartina*, but equally well, or better developed air spaces occur in roots of sugar cane (Plate 1), and of *Paspalum notatum*, and in some roots of maize (Plate 1). Not only does this aerenchyma provide an internal route of aeration, but it may also provide a larger surface for bacterial colonization than occurs on the surface of the root. It may be pertinent that aerobic diazotrophs have been implicated as responsible for nitrogenase activity associated with sugar cane and *Paspalum*,[29] both normally terrestrial grasses with a well developed aerenchyma. Nonobligate microaerophils are associated with other grasses with less well developed aerenchyma or none, and an obligate microaerophil is associated with *Spartina*, an aquatic plant growing in sulfidic sediments, i.e., there is some suggestion of a correlation of microbial respiratory metabolism with the efficiency of aeration of the roots. The association of a facultative anaerobe (*Bacillus*) with wheat roots, as reported in this Workshop, appears to be consistent with this scheme as roots of wheat have a very compact cortex. Well developed hypodermal layers occur in some roots (Plate 1), and must restrict lateral gas exchange. In combination with the air space system, this may result in fairly stable internal oxygen concentrations. Almost certainly, bacterial colonization is restricted by this hypodermis, and this is probably one reason why initial colonization occurs at root apices, or where lateral roots break through the cortex.[10,11]

The possible role of the large nondiazotrophic component of the endorhizosphere population on infection and expression of nitrogenase activity by the diazotrophs needs to be investigated. It is almost inconceivable that there are not critical interactions of some sort between the different groups of microorganisms in the roots. A possible analogy is the facilitation of growth of normally saprophytic bacteria within roots by the presence of a pathogen.[31]

Finally, we suggest that in spite of the denigration it has received, excised root nitrogenase activity does have some validity as a parameter (not a direct measure) of root nitrogenase activity, as all of the results we have obtained using intact plant systems were predictable from the results obtained using excised root systems. Preincubation of roots with ammonium and addition of $HgCl_2$ after a preincubation period may be an appropriate modification of the traditional methodology which would allow screening of plants for internal and possibly "coupled" root-diazotroph systems.

IX. POSTSCRIPT: C_2H_2 ASSAYS

Recently it has been demonstrated using $^{14}C_2H_2$ that in some systems of low apparent acetylene-reducing activity, a substantial part of the C_2H_4 produced in the presence of C_2H_2 is not derived from C_2H_2 reduction.[32] As C_2H_2 blocks oxidation of C_2H_4, controls lacking C_2H_2, do not necessarily indicate endogenous levels of C_2H_4 production.

We have found no C_2H_4 production in excised root systems, with or without added glucose, when incubated under N_2, i.e., in which there would be no oxidation of any C_2H_4 produced. The absence of C_2H_4 production may be associated with the marine nature of this system. The kinetics of (apparent) acetylene reduction in *Spartina* excised root systems are similar to those for pure cultures of diazotrophs.[14] Van Berkum and Sloger[21] also tested for oxidation of added C_2H_4 by roots of *Spartina* and found very low rates (0.56 nmol C_2H_4 oxidized per gram dry root per hour) in comparison to the C_2H_4 production in the presence of C_2H_2.

REFERENCES

1. **Mann, K. H.,** in *Estuarine Research,* Vol. 1, Cronin, L. E., Ed., Academic Press, New York, 1975, 634.
2. **Patriquin, D. G.,** *Ecol. Bull. Stockholm,* 26, 20, 1978.
3. **Döbereiner, J. and Day, J. M.,** in *Proceedings 1st International Symposium on Nitrogen Fixation,* Newton, W. E. and Nyman, C. J., Eds., Washington State University Press, Pullman, 1976, 518.
4. **Tarrand, J. J., Krieg, N. R., and Döbereiner, J.,** *Can. J. Microbiol.,* 24, 967, 1978.
5. **Patriquin, D. G. and McClung, C. R.,** *Mar. Biol.,* 47, 227, 1978.
6. **Smibert, R. M.,** *Annu. Rev. Microbiol.,* 32, 673, 1978.
7. **Pedersen, W. L., Chakrabarty, K., Klucas, R. V., and Vidaver, A. K.,** *Appl. Environ. Microbiol.,* 35, 129, 1978.
8. **Okon, Y., Albrecht, S. L., and Burris, R. H.,** *Appl. Environ. Microbiol.,* 33, 85, 1977.
9. **Patriquin, D. G., Ruschel, A. P., and Graciolli, L. A.,** in UNDP/IAEA Application of Nuclear Technology in Agriculture, Brasil, Plant Biochemistry: A Consultancy Report, BRA/71/556, Tech. Rep. 16, International Atomic Energy Agency, Vienna, 1978.
10. **Patriquin, D. G. and Döbereiner, J.,** *Can. J. Microbiol.,* 24, 734, 1978.
11. **Magalhães, F. M. M., Patriquin, D., and Döbereiner, J.,** *J. Acad. Bras. Cien.,* in press.
12. **Lakshmi, V., Rao, A. S., Vijayalakshmi, K., Kumari, M., Tilak, K., and Subba Rao, N. S.,** *Proc. Indian Acad. Sci.,* 86B, 397, 1977.
13. **Bidwell, R. G. S.,** *Plant Physiology,* Macmillan, New York, 1972.
14. **Patriquin, D. G.,** *Aquat. Bot.,* 4, 193, 1978.
15. **Patriquin, D. G. and Denike, D.,** *Aquat. Bot.,* P. 211, 1978.
16. **Boyle, C. D.,** Some Characteristics of Nitrogenase Activity Associated with Diazotrophs in and around *Spartina alterniflora* Roots, Master's thesis, Dalhousie University, Halifax, N.S., Canada, 1979.
17. **Stewart, W. D. P. and Rowell, P.,** *Nature London,* 265, 371, 1977.
18. **Campbell, N. E. R. and Evans, H. J.,** *Can. J. Microbiol.,* 15, 1342, 1969.
19. **Döbereiner, J., Day, J. M., and Dart, P. J.,** *J. Gen. Microbiol.,* 71, 103, 1972.
20. **Teal, J. M. and Kanwisher, J. W.,** *J. Exp. Bot.,* 17, 355, 1966.
21. **van Berkum, P. and Sloger, C.,** *Plant Physiol.,* in press.
22. **Livingstone, D. C.,** Growth and Nitrogen Accumulation in *Spartina alterniflora* Loisel, in Relation to Certain Environmental and Physiological Factors, Master's thesis, Dalhousie University, Halifax, N. S., Canada, 1978.

23. **McManmon, M. and Crawford, R. M. M.,** *New Phytol.,* 70, 299, 1971.
24. **Okon, Y., Albrecht, S., and Burris, R. H.,** *J. Bacteriol.,* 128, 592, 1976.
25. **Raven, J. A. and Smith, F. A.,** *New Phytol.,* 76, 415, 1976.
26. **Christeller, J. T., Laing, W. A., and Sutton, W. D.,** *Plant Physiol.,* 60, 47, 1977.
27. **Teal, J. M., Valiela, I., and Verlo, D.,** *Limnol. Oceanogr.,* 24, 126, 1979.
28. **Smith, D.,** in *The Biology and Utilization of Grasses,* Younger, V. B. and McKell, C. M., Eds., Academic Press, New York, 1974, 318.
29. **Neyra, C. A. and Döbereiner, J.,** *Adv. Agron.,* 29, 1, 1978.
30. **Evans, L. T. and Wardlaw, I. F.,** *Adv. Agron.,* 28, 301, 1976.
31. **Young, J. M. and Patton, A. M.,** in Proc. 3rd Int. Conf. on Plant Pathogenic Bacteria, Wageningen, Geesteranius, H. P. M., Ed., Ctr. Agric. Publ. Document (PUDOC), Wageningen, Netherlands, 1972, 77.
32. **Witty, J. F.,** *Soil Biol. Biochem.,* 11, 209, 1979.
33. **McClung, C. R. and Patriquin, D. G.,** unpublished material.
34. **Boyle, C. D.,** unpublished observation.
35. **Valiela, Ivan and Smith, David,** personal communication.

Chapter 3

ACETYLENE REDUCTION ACTIVITY (ARA) BY ENDORHIZOSPHERE DIAZOTROPHS

D. G. Patriquin* and C. D. Boyle**

TABLE OF CONTENTS

*   Biology Department, Dalhousie University, Halifax, N.S., Canada.
** Programa Fixacao Biologica de Nitrogenio, Rio de Janeiro, Brazil.

I. INTRODUCTION

Acetylene dependent ethylene production (ARA) was measured separately around the roots and leaves of intact hydroponically grown *Spartina alternifolia* and rice (dryland cv. IAC-25) in two compartment assay chambers. Upper phase C_2H_4 could originate from: (1) ARA of diazotrophs in the leaves, (2) diffusion of the C_2H_4 from around the roots in the lower phase (via the plants' air space systems) into the upper phase, or (3) ARA of diazotrophs in the roots (endorhizosphere), the resultant C_2H_4 diffusing via the air space system into the upper phase. There is evidence for the existence of endorhizosphere diazotrophs for both *S. alterniflora*[3] and rice,[1,4] but it is not clear whether these are active in the plant.

II. EXPERIMENTAL

A. Plant Assay System

Plants were transferred from a hydroponics tank to the apparatus shown in Figure 1, and on the following day the tubes were sealed over the leaves. C_2H_2 was introduced through the lower phase sampling part to displace water through the vent for an 8% equilibrium C_2H_2 concentration. Eight percent C_2H_2 was injected into the upper phase. After 1 and 5 hr, 0.5 mℓ gas samples were withdrawn and analyzed by gas chromatography. The systems were shaken after C_2H_2 addition and before sampling.

B. Determination of the Effect of HgCl$_2$ Surface Sterilization on ARA

Roots, rhizomes, or soils + 0.1% glucose were preincubated at 3 to 5% O_2 in flasks for 15 hr and 10% C_2H_2 and 0.2% (*S. alterniflora*) or 0.1% (rice) HgCl$_2$ were then injected. Gas samples of 0.5 mℓ were withdrawn and analyzed for C_2H_4 after 2 and 4 hr (*S. alterniflora*) or 1 and 3 hr (rice). For *S. alterniflora,* young roots were cut from the periphery of the root mass of plants which had been transplanted from the field to a bucket 8 months previously. Older roots and rhizomes were cut from within this root mass. All of the roots from field grown rice plants at mid seed filling stage were collectively assayed.

III. RESULTS AND DISCUSSION

The relative contribution of source one to the upper phase activity of *S. alterniflora* was determined by measuring the rate of ethylene appearance in the upper phase on two successive days, and comparing this to the ARA of leaves excised from these plants on the following day. Leaf associated ARA accounted for approximately 16% of the upper phase ethylene appearance (Table 1).

The expected flux of ethylene from around the roots into the upper phase (source 2) was calculated as described by Boyle and Patriquin.[2] For *S. alterniflora,* this flux (F C_2H_4) was usually much lower than the upper phase rate of ethylene appearance. In contrast, source two accounted for essentially all of the upper phase activity of rice (Table 2).

It is proposed that the residual upper phase ethylene appearance, that not accounted for by leaf associated ARA or diffusion from around the roots, is the product of endorhizosphere diazotrophs' nitrogenase activity. In the experiments presented here this activity was considerably greater for *S. alterniflora* than for rice.

In support of this proposal, a part of the ARA of some roots excised from *S. alterniflora* persisted after surface sterilization, while ARA of rice roots was eliminated (Table 3). The method of sterilization employed with *S. alterniflora* (0.2% HgCl$_2$

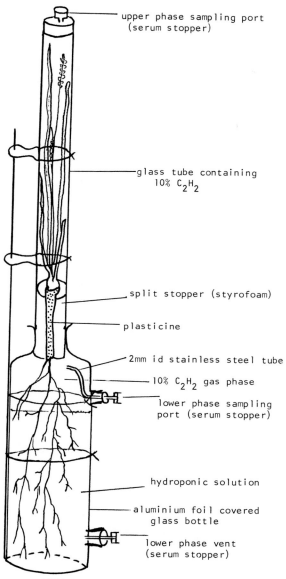

upper phase sampling port
(serum stopper)

glass tube containing
10% C_2H_2

split stopper (styrofoam)

plasticine

2mm id stainless steel tube

10% C_2H_2 gas phase

lower phase sampling
port (serum stopper)

hydroponic solution

aluminium foil covered
glass bottle

lower phase vent
(serum stopper)

FIGURE 1.   The plant assay system.

Table 1
RATES OF ETHYLENE APPEARANCE IN UPPER PHASES AND ARA OF CORRESPONDING EXCISED LEAVES FOR *S. ALTERNIFLORA*

Plant	A Upper Phase	B Excised Leaves	B as a % of A
1	12.6 ± 2.0[a]	1.7[b]	13%
2	8.8 ± 2.8	1.7	19%
3	6.7 (one assay)	1.6	24%
4	13.9 ± 4.4	1.1	8%
			Average = 16%

[a]   Mean ± range for two consecutive assays. Leaf tubes were removed and the water was replaced after each assay.

[b]   nmol C_2H_4/hr.

Table 2
COMPARISON OF THE RATE OF C₂H₄ APPEARANCE IN THE UPPER PHASES OF PLANT ASSAY SYSTEMS TO THAT EXPECTED FROM DIFFUSION FROM THE LOWER PHASE

Plant	Observed rate[a]	Calculated rate
S. alterniflora	18.3	0.6
	14.9	2.1
	4.2	2.1
	16.8	1.4
	26.0	0.5
	15.6	0.6
Rice	5.8	4.1
	9.4	6.4
	7.9	4.6
	5.0	2.6
	4.6	4.0
	2.6	4.0

[a]   nmol C_2H_4/hr.

Table 3
THE EFFECT OF HGC1₂ ADDITION ON ARA OF EXCISED ROOTS, RHIZOMES, OR RHIZOSPHERE SOILS FROM *S. ALTERNIFLORA* OR RICE

System	ARA[a]		B as a % of A
	A Controls	B + HgCl₂	
S. alterniflora			
Young roots	40 ± 9.6	1.7 ± 0.5	4.2
Older roots	39 ± 7.4	10.7 ± 2.3	27.4
Rhizomes	74 (one sample)	3.3	4.4
Soils	431 ± 8.2	0.1	0
Rice			
Roots	10.2 ± 2.0	0	0
Soils	415 ± 10	6 ± 2	1

[a]   nmol C_2H_4/hr/wet g; mean ± range for 2 replicates.

added 2 hr before start of assay) eliminated all activity of glucose amended rhizosphere soil, external diazotroph population, while that used with rice (0.1% $HgCl_2$ added 1 hr before the start of assay) left a small residual soil activity.

It is concluded that *S. alterniflora* has an active internal diazotroph population, while that of the cultivar of rice used here is relatively inactive.

REFERENCES

1. **Baldani, V. L. D. and Döbereiner, J.,** Host plant specificity in the infection of cereals with *Azospirillum* spp., *Soil Biol. Biochem.*, 1979 in press.
2. **Boyle, C. D. and Patriquin, D. G.,** in preparation.
3. **Patriquin, D. G. and McClung, C. R.,** Nitrogen accretion, and the possible significance of N_2-fixation (acetylene reduction) in a Nova Scotian *Spartina alterniflora* stand, *Mar. Biol.*, 47, 277, 1978.
4. **Watanabe, I. and Barraquio, W. L.,** Low levels of fixed nitrogen required for the isolation of free-living N_2-fixing organisms from rice roots, *Nature (London)*, 277, 565, 1979.

Chapter 4

INCREASE OF ANTIBIOTIC RESISTANT BACTERIA IN GRASS ROOTS

Vera Lucia D. Baldani* and Johanna Döbereiner**

TABLE OF CONTENTS

*   Graduate student of the Universidade Federal Rural do Rio de Janeiro.
** Programa Fixacao Biologica de Nitrogenio, Rio de Janeiro, Brazil.

I. INTRODUCTION

Crop rotation with cereal crops has been recommended as biological control for certain bacterial diseases for many years[7] and the stimulation of actinomycetes[3,8] and enrichment of antibiotics in the rhizosphere[5] are well known. Still, little attention has been given to this apparently powerful tool to manipulate the microbial equilibrium in the rhizosphere. Brown[2] reported enrichment of streptomycin resistant microorganisms in the rhizosphere of wheat, legumes, and vegetables. Streptomycin resistance has been used as a marker in many *Rhizobium* inoculation experiments but no selective advantage has been reported under field conditions.[4]

Recently, host plant specificity in the infection of maize, rice, and wheat by *Azospirillum* spp. has been demonstrated,[1] where maize was selected for *Azospirillum lipoferum* and wheat and rice for the *A. brasilense* nir⁻ group. To confirm the specific infection by *Azospirillum*, strains marked with streptomycin resistance were used as inoculants in a field experiment with maize and wheat. In both species, in the noninoculated control plots and in the plots inoculated with heterologous strains, a high proportion of homologous streptomycin resistant *Azospirillum* spp. were obtained from surface sterilized roots but not from rhizosphere soil. These observations indicated selection not only for certain strains but also for streptomycin resistance, during the infection of the roots.

To confirm these observations the proportion of streptomycin resistant bacteria in soil was compared with that in the rhizosphere and roots of maize and sorghum collected in the field.

II. METHODS

A. Incidence of Spontaneous Streptomycin Resistant *Azospirillum lipoferum* in Rhizosphere Soil and Surface Sterilized Maize or Wheat Roots and the Effect of Inoculation

Inoculation consisted of application of 150 mℓ/m^2 of a liquid culture of *A. lipoferum* str^r or *A. brasilense* str^r, strains isolated from surface sterilized maize or wheat roots, respectively, and marked with streptomycin resistance (20 μg/mℓ). The percentage of bacteria tolerant to streptomycin was calculated from numbers of colonies on potato agar plates with and without 20 μg/mℓ of streptomycin. Each point in Figures 1 and 2 was calculated from six to eight isolates. The soil contained about 10^{-8} spontaneous streptomycin resistant *Azospirillum* spp. before planting.

Surface sterilized roots were treated for 30 min with a 1% solution of chloramine-T, which kills most microorganisms on the root surface but leaves bacteria in the stele and inner cortex alive.[6] Rhizosphere soil was treated with streptomycin by applying 1 ℓ of a 4000 ppm solution of "Distreptine" over each maize plant 7 and 2 days before sampling. Distreptine is a commercial product containing streptomycin for treatment of bacterial diseases in tomatoes.

B. Enrichment of Streptomycin and Penicillin Resistant Bacteria in Field Grown Sorghum

Sorghum was grow with four treatments (20 ton compost/per hectare and 40 kg NO$_3$-N/ha in all combinations) and four replicates in a factorial field experiment. Composite soil samples were taken between rows. Roots of two plants/sample were removed from the field, washed and sterilized intact for 30 min in 1% chloramine-T, 10 g roots were disintegrated in a star mixer with 100 mℓ of 4% sucrose solution and serial dilutions were plated on potato agar. The percentage of resistant bacteria was calculated from colony numbers on plates with and without 10 μg/mℓ of streptomycin or 10 IU/mℓ of penicillin.

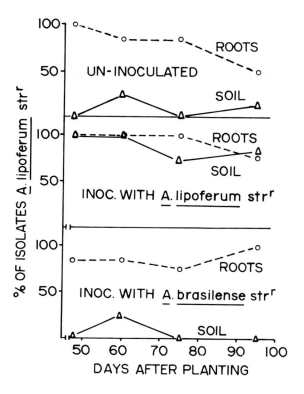

FIGURE 1. Incidence of spontaneous streptomycin resistant *Azospirillum lipoferum* in rhizosphere soil and surface sterilized (1 hr) roots of maize grown in the field, and effect of inoculation. In the uninoculated control plots, 84% from the rhizosphere soil isolates and 96% from the root isolates were *A. lipoferum*.

III. RESULTS

Compared with soil, the percentage of resistant bacteria in the rhizosphere of maize increased more than 10 times, and that in washed and surface sterilized roots more than 1000 times (Table 1). Surface sterilizations of roots increased the incidence of streptomycin resistant bacteria when compared with washed roots. The percentage of streptomycin resistant bacteria in sorghum roots (Table 2) was lower, but still substantially higher than in soil. In addition there were higher proportions of penicillin resistant bacteria in roots than in soil (Table 2). These experiments indicate the requirement of certain resistance to antibiotics especially to streptomycin for infection of cereal roots. *Azospirillum* strains isolated from surface sterilized roots of field grown maize and wheat plants were predominantly streptomycin resistant while most isolates from unsterilized roots and soil were inhibited by 20 $\mu g/m\ell$ (Figures 1 and 2). The high proportion of resistant *A. lipoferum* in roots of maize and *A. brasilense* nir$^-$ in roots of wheat persisted during the whole growth cycle and was independent of inoculation treatments. The importance of this apparent selection mechanism of cereal plants for certain microorganisms is stressed, with new possibilities for the introduction of selected or manipulated nitrogen fixing bacteria.

Some more detailed results of this paper are given in *Can. J. Microbiol.*, 1979.

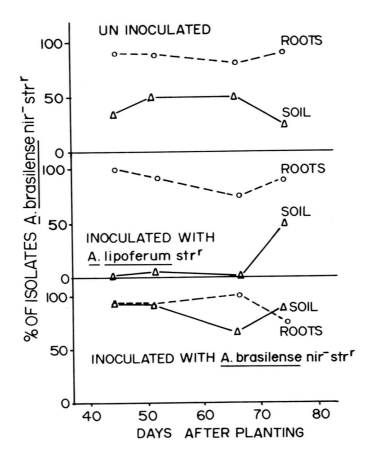

FIGURE 2.   Incidence of spontaneous streptomycin resistant *Azospirillum brasilense* nir⁻ in rhizosphere soil and surface sterilized (15 min) roots of wheat grown in the field, and effect of inoculation. In the uninoculated control plots, 43% from the rhizosphere soil isolates and 88% from the root isolates were *A. brasilense* nir⁻.

Table 1
ENRICHMENT OF STREPTOMYCIN RESISTANT BACTERIA IN MAIZE ROOTS[3]

Sample	% of bacteria tolerant to 20 μg/ml of streptomycin		
Soil collected between rows	0.06	±	0.1
Rhizosphere soil	0.72	±	0.4
Surface sterilized roots	96.60	±	14.6
Rhizosphere soil treated with Distreptine	2.6	±	0.2

Note: Values are means of 3 samples each of two soils with standard deviations of the means.

Table 2
ENRICHMENT OF STREPTOMYCIN (10 μg/ml) AND PENICILLIN (10 IU/ml) RESISTANT BACTERIA IN ROOTS OF FIELD GROWN *SORGHUM*

Soil treatment	Total no. of bacteria $\times 10^7$		% resistant to streptomycin		% resistant to penicillin	
	Soil	Root	Soil	Root	Soil	Root
Organic matter + N	190	522	0.16	47	0.23	145
Organic matter	113	163	0.20	36	0.42	71
N	31	367	1.30	35	1.80	99
Check	49	128	0.50	24	1.01	76

Note: Values are means of two plots, two plants each.

REFERENCES

1. **Baldani, V. L. D. and Döbereiner, J.,** Host plant specificity in the infection of cereal with *Azospirillum* spp., *Soil Biol. Biochem.,* 1979, in press.
2. **Brown, M. E.,** Stimulation of streptomycin resistant bacteria in the rhizosphere of leguminous plants, *J. Gen. Microbiol.,* 24, 369, 1961.
3. **Freitas, J. L. M., Pereira, P. A. A., and Döbereiner, J.,** Effect of Organic Matter and *Azospirillum* Strain on N Metabolism in *sorghum vulgarae,* Int. Workshop on Associative N$_2$-Fixation, Piracicaba, Brazil, 1979.
4. **Gareth-Jones, D. and Bromfield, E. S. P.,** A study of the competitive ability of streptomycin and spectinomycin mutants of *Rhizobium trifolii* using various marker techniques, *Ann. Appl. Biol.,* 88, 448, 1978.
5. **Krasil'nikov. N. A.,** Soil Microorganisms and Higher Plants, Acad. Sci. U.S.S.R. (trans.), National Science Foundation and Department of Agriculture, Washington, D.C., 1958.
6. **Magalhães, F. M. M., Patriquin, D., and Döbereiner, J.,** Infection of maize roots by *Azospirillum* spp., Associative N$_2$-Fixation, Vose, P. B. and Ruschel, A. P., Eds., CRC Press, Boca Raton, Fla., 1980.
7. **Robbs, C. F.,** Bacterioses Fitopatogenicas no Brasil, Ins. Econ. Rural, 1960.
8. **Rovira, A. D.,** Interactions between plant roots and soil microorganisms, *Annu. Rev. Microbiol.,* 19, 241, 1965.

Chapter 5

NITROGEN FIXING BACTERIA ISOLATED FROM DIVERSE SOILS AND GRASS ROOTS IN AMAZONIA

Fatima Maria Moreira Magalhães*

TABLE OF CONTENTS

* Division of Agronomy, Instituto Nacional de Pesquisas da Amazonia Manaus, Am., Brazil.

Table 1

TOTAL NUMBERS OF SAMPLE (LOOPFULS OF SOIL AND PIECES OF ROOT) AND NUMBER OF SAMPLES POSITIVE FOR N-FIXATION FOR EACH CULTURE MEDIUM AND PLACE OF SAMPLE COLLECTION

Medium Carbon sources	NFb semisolid (malate)		LG mod. semisolid (sucrose + malate)		Becking (Beij.) liquid (glucose)		Becking (Azot.) liquid (glucose)		Lipman sol. (amid)		Becking (Beij.) sol. (glucose)		Mixed semisol. (malate + mannitol + sucrose)		LG + CaCO₃ sol. (sucrose)		% total of positive samples[c]
	a	b	a	b	a	b	a	b	a	b	a	b	a	b	a	b	
Varzea Manaus (roots and soils)	28	27	28	27	28	15	29	10	120	6	120	5	—		—		34%
Terra-firme Manaus (roots and soils)	24	16	24	17	—		—		96	4	96	0	24	19	24	8	25.7%
Varzea Belem (roots)	15	15	15	15	—		—		28	24	—		15	15	60	45	93.1%
Terra-firme Belem (roots)	21	4	16	8	—		—		84	34	—		20	10	84	34	38%

Note: a = total numbers of samples.

b = numbers of positive samples, (showing acetylene reduction in culture medium).

c = calculated from total samples grown in media NFb s.s., LG modif.s.s., and Lipman solid.

Table 2
SITES, PLANTS, AND MEDIUMS FROM WHICH NITROGEN-FIXING BACTERIA WERE ISOLATED

Code	Probable species or genus	Plant	Site	Medium
NST4P1	*Azospirillum lipoferum*	*Paspalum plicatum* (soil)	Terra firma	NFG semi-solid
15M1	*Azospirillum* spp.	*Brachiaria brizantha* (roots)	Terra firma	Semisolid mixture
5'A3	*Derxia* spp.	*Oriza perenne* (roots)	Várzea	Lipman (Mod.)
L + CRT3 P1	*Azotobacter* spp.	Quicuio (roots)	Terra firma	LG + CaCO₃ solid
LRT4P2	*Azotobacter* spp.	*Paspalum plicatum* (roots)	Terra firma	LG Mod. semisolid
17S	*Azotobacter* spp.	*Sorghum arundinaceum* (soil)	Várzea	Becking (*Azot.*) solid
7'A	*Azotobacter* spp.	*Sorghum arundinaceum* (roots)	Várzea	Becking (*Azot.*) solid
10	?	Canarana (roots)	Várzea	Lipman (Mod.)
4'	?	Canarana (roots)	Várzea	Lipman (Mod.)

percentage (93%) occurred in flooded soil (Varzea-Belem). This result may seem surprising given that the bacteria concerned are aerobic. However, Döbereiner and Campelo[10] made the same kind of observation on the occurrence of *Derxia* spp., and also Döbereiner and Alvahydo[9] in relation to the occurrence of *Beijerinckia* spp. Schmidt and Rippel[14] explained this apparent paradox by suggesting that nitrogen fixation, a reductive process, may be inhibited by the high pO_2 of the atmosphere.

Nitrogen fixing organisms representing the genus *Azotobacter, Beijerinckia, Derxia,* and *Azospirillum* were commonly found in the Amazon region. There is an apparent preferential distribution of these organisms in relation to the area studied. *Azospirillum* and *Azotobacter* occurred in samples from the four areas of collection. *Derxia* was only isolated from the flooded Varzea of Belem, and *Beijerinckia* was detected only in the terra firma of Belem. No *Derxia* and *Beijerinckia* species were identified in samples from the two collection sites of Manaus. Several other nitrogen fixing organisms were isolated in pure culture whose identification is underway in our laboratory.

B. Characteristics of Nine Isolated Diazotrophs

Table 2 shows places, plants, and medium from which nine nitrogen-fixing bacteria were isolated. Morphological characteristics of their colonies in both solid potato and in GNA media, as well as the catalase activities are described in Table 3. Results of N_2-fixing ability, i.e., nitrogenase activity, of each bacteria with various carbon sources, are shown in Table 4.

The criterion for positive nitrogenase activity (C_2H_2 reduction) was an ethylene peak at least five times greater than controls. All bacteria tested grew on sucrose and glucose. In general these media also gave highest growth and nitrogenase activity. The most selective carbon sources were rhamnose and xylose, since they supported growth of only two and three bacteria, respectively.

Table 3
GENERAL CHARACTERISTICS OF DIAZOTROPHS ISOLATED FROM AMAZON REGION

Code	Catalase	GNA medium	Potato medium
LRT4P2	+	Hard, amber, better growth on GNA medium than on potato medium. Color and consistency are kept with age, wrinkled surface and irregular edges	As GNA, but less growth
L + C RT3P1	+	As LRT4P2	As LRT4P2
NST4P1	−	Typical *Azospirillum* colonies, dry, cream coloured, firm consistency. Characteristics are kept with age	
5′A3	−	Wrinkled surface, brown, extremely hard and elastic, same aspect on solid LG medium, with age the consistency becomes more fluid	Much less growth than on GNA, elastic consistency and smooth surface Characteristics are kept with age
15 M1	+	White, ± firm consistency, edges and surface irregular	Smaller than on GNA, edges irregular
17 S	+	Dense, ± firm consistency, cream colored, smooth surface, with age the consistency becomes more fluid	Smaller than on GNA, irregular edges Characteristics are kept with age
7′a	+	As 17 S	As 17 S
10	−	Better growth than on potato medium; dark yellow colonies	Yellow, irregular edges, wrinkled surface, ± firm consistency
4′	−		White, ± firm consistency, irregular edges, wrinkled surface, with age colonies become flatter

Plates 1 and 2 show phase contrast microscopy photographs of the nine isolates studied above, and representatives of the genus *Beijerinckia* and *Azotobacter*. On a basis of the characteristics studied, probable genera or species were assigned to these bacteria (Table 2). Two bacteria were positively identified (*Azospirillum lipoferum*-NST4P1, and *Derxia* spp.-5′A3); of the others four probably belong to the genus *Azotobacter,* one to the genus *Azospirillum* and two remained unidentified since they did not correspond to any description found in the literature.

In the case of organisms assigned to genera but not identified at species level, there were divergences between the characteristics observed in this study and those reported in the literature. However, the relatively small number of characteristics studied did not permit determination of whether these differences represent new genera or species. The characteristics studied will help to identify these bacteria in future work on the occurrence, isolation, and identification of N₂-fixing bacteria in Amazonian systems.

ACKNOWLEDGMENT

Thanks are due to Dr. Rosemary Sylvester-Bradley for general guidance in this work.

Table 4

NITROGENASE ACTIVITY ON DIFFERENT CARBON SOURCES

Code	Probable species or genus	Malate	Mannitol	Amide	Saccharose	Glycerol	Glucose	Lysine	Arginine	Succinate	Arabinose	Raffinose	Lactose	Rhamnose	Xylose	Galactose
NST4P1	*Azospirillum lipoferum*	+	+	+	+	+	+	+	–	+	+	–	+	–	+	+
15M1	*Azospirillum* spp.	+	+	+	+	+	+	+	–	+	+	–	+	+	+	+
5'A3	*Derxia* spp.	+	+	+	+	+	+	+	+	+	–	–	–	–	–	–
L + CRT3P1	*Azotobacter* spp.	+	+	+	+	+	+	+	+	+	+	–	–	–	–	–
LRT4P2	*Azotobacter* spp.	–	+	+	+	–	+	–	ND[a]	–	+	+	+	+	–	–
17S	*Azotobacter* spp.	+	+	–	+	–	+	–	ND[a]	+	–	+	–	–	–	+
7'a	*Azotobacter* spp.	+	–	–	+	+	+	–	–	+	–	+	–	–	–	+
10	?	–	–	–	+	+	+	–	–	–	+	–	–	–	–	–
4'	?	+	+	+	+	+	+	–	–	+	–	+	–	–	+	–

[a]   ND — Not determined.

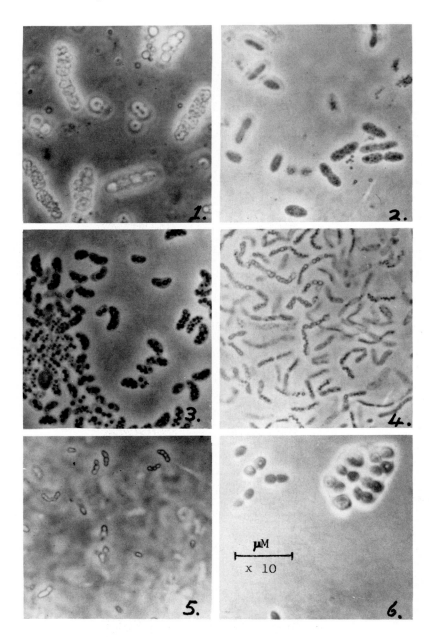

PLATE 1. Phase contrast microscopy of diazotrophs isolated from the Amazon region. (1) LRT4P2 (*Azospirillum* sp.), pure culture on semisolid; sucrose medium; (2) L + C RT3P1 (*Azospirillum* sp.), pure culture on semisolid mixed; (3) NST4P1 (*Azospirillum lipoferum*), pure culture on semisolid glucose; (4) 5′A3 (*Derxia* sp.), pure culture on malate semisolid; (5) 15M1 (*Azospirillum* sp.), pure culture on semisolid sucrose; (6) 17 S (*Azotobacter* sp.), pure culture on semisolid mixed.

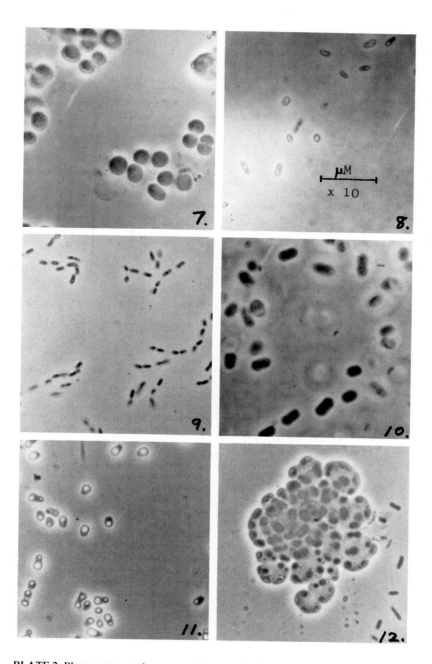

PLATE 2. Phase contrast microscopy of diazotrophs isolated from the Amazon region. (7) 7′a (*Azotobacter* sp.), pure culture on semisolid mixed; (8) 4′ Pure culture on semisolid mixed; (9) 10 Pure culture on semisolid mixed; (10) (*Azotobacter* sp.), culture on liquid starch ex root of *Panicum maximum;* (11) (*Beijerinckia* sp.), enrichment culture ex root of *Hyparrhenia rufa*, on solid Lipman mod; (12) (*Beijerinckia fluminense*), enrichment culture ex root of *Hyparrhenia rufa*, on solid Lipman mod.

REFERENCES

1. **Asakawa, N. M.,** Master's Thesis, Universidade do Amazonas, Manaus, Brazil, 1978.
2. **Becking, J. H.,** *Plant Soil,* 14, 49, 1961.
3. **von Bülow, J. F. W. and Döbereiner, J.,** *Proc. Nat. Acad. Sci., U.S.A.,* 53, 532, 1975.
4. **Campelo, A. B. and Döbereiner, J.,** *Pesqui. Agropec. Bras.,* 5, 327, 1970.
5. **Day, J. M. and Döbereiner, J.,** *Soil Biol. Biochem.,* 8, 45, 1976.
6. **de-Polli, H., Matsui, E., Döbereiner, J., and Salati, E.,** *Soil Biol. Biochem.,* 8, 1, 1976.
7. **Döbereiner, J.,** *Pesqui. Agropec. Bras.,* 1, 357, 1966.
8. **Döbereiner, J.,** personal communication, 1979.
9. **Döbereiner, J. and Alvahydo, R.,** *Ciênc. Cult. Sao Paulo,* 11, 208, 1959.
10. **Döbereiner, J. and Campelo, A. B.,** *Plant Soil,* Special Vol., 457, 1971.
11. **Döbereiner, J., Dart, S. M., and Dart, P. J.,** *Plant Soil,* 37, 191, 1972.
12. **Ruschel, A. P.,** Escola Superior de Agricultura "Luiz de Queiroz", Doctoral thesis, University of São Paulo, Brasil, 1976.
13. **Ruschel, A. P., Henis, Y., and Salati, E.,** *Soil Biol. Biochem.,* 7, 181, 1975.
14. **Schmidt, L. W. and Rippel, A. B.,** *Arch. Microbiol.,* 28, 45, 1957.
15. **Neyra, C. A. and Döbereiner, J.,** *Adv. Agron.,* 29, 1, 1977.
16. **Sylvester-Bradley, Rosemary,** personal communication.

Chapter 6

A NOTE ON NITROGENASE AND NITRATE REDUCTASE ACTIVITIES, AND DENITRIFICATION IN FIVE *BRACHIARIA* SPP. GENOTYPES*

Pedro A. A. Pereira and Johanna Dobereiner**

TABLE OF CONTENTS

*   Translator: Diva Athié.
** Programa Fixação Biologica de Nitrogênio, Rio de Janeiro, Brazil.

I. INTRODUCTION

Research on biological N_2-fixation in *Gramineae* has greatly increased in the last 3 years. The objective of such research, using different methods, is to find out the potential for biological fixation of atmospheric N_2 by grasses and its application in Tropical Agriculture.

There are, however, many limiting factors to biological nitrogen fixation in the tropics, because nitrogenase activity in the field is a result of the interaction of ecological and physiological factors and plant-bacteria association.

On the other hand, genetic work with *Gramineae* generally takes into consideration plants whose genotype responds better to nitrogen fertilization. However, it has been shown that there is a variation between cultivars of *Paspalum notatum, Pennisetum purpureum* and *Digitaria decumbens*[1,2] in relation to nitrogenase activity, suggesting the possibility of selecting genotypes more efficient in nitrogen fixation.

Data obtained in previous work has proved that during the vegetative growth period (which is rather quick) of forage grasses, there is a considerable fixation of nitrogen, while at flowering and especially during the cool and dry season, there is very little nitrogen fixation.[3]

II. EXPERIMENTAL

Due to the importance of improving pastures to make them self-sufficient in nitrogen through biological fixation, an experiment was carried out with five genotypes of *Brachiaria*.

It was noted that *Brachiaria ruziziensis* CPI-30623 showed nitrogenase activity significantly higher than the other four genotypes. In all genotypes nitrogenase activity was high during the period of highest temperature and humidity, which coincided with plant vegetative growth, and the smallest activity was found during the cool and dry period, exactly at flowering of all genotypes. In all genotypes nitrogen fertilization (3 × 20 kg N/ha) had no significant effect on nitrogenase activity.

The genotype which showed the highest capacity for NO_3^- reduction in the leaves was *Brachiaria radicans*. All the other genotypes showed a low nitrate reductase activity, even when NO_3^- was applied 48 hr before sampling.

In the present work, comparison was also made of nitrogenase activity using two different methods: excised roots and undisturbed soil-plant system. When comparing actual nitrogen harvested in three forage cuttings with estimates from the acetylene reduction method, it was noticed that the method can underestimate nitrogenase activity in the field, suggesting that the results obtained using C_2H_2 reduction in *Gramineae* is only of a qualitative value.

Denitrification was also measured (by N_2O accumulation under C_2H_2[4]) in the five *Brachiaria* spp. genotypes. Sixty-nine hours after nitrate application up to 7% of the fertilizer had been denitrified. Many *Azospirillum* spp. strains are known to be denitrifiers and they are possibly among those responsible for this loss.

REFERENCES

1. **Döbereiner, J. and Day, J. M.,** Nitrogen fixation in the rhizosphere of tropical grasses, in Nitrogen Fixation by Biosphere, IBP Synthesis Meeting, Vol. 1, Stewart, W. D. P., Ed., Edinburgh, 1975, 39.
2. **Döbereiner, J. and Day, J. M.,** Associative symbiosis in tropical grasses, characterization of microorganisms and dinitrogen fixing sites, in *International Symposium on Nitrogen Fixation*, Newton, W. E. and Nyman, C. J., Eds., 1976. Washington State University Press, Pullman, 1976.
3. **Neves, M. C. P., Day, J. M., Carneiro, A. M., and Döbereiner, J.,** Atividade da nitrogenase na rizosfera de gramíneas tropicais forrageiras, *Rev. Microbiol.*, 7(3), 1976.
4. **Yoshinari, T. and Knowles, R.,** Acetylene inhibition of reduction of nitrous oxide by denitrifying bacteria, *Biochem. Biophys. Res. Commun.*, 69, 705, 1976.

Chapter 7

CONTRIBUTION TO THE METHODOLOGY OF EXCISED ROOT ASSAYS TO EVALUATE NITROGENASE ACTIVITY IN *GRAMINEAE**

J. I. Baldani, P. A. A. Pereira, C. A. Neyra, and J. Dobereiner**

TABLE OF CONTENTS

* Translator: Diva Athie.
** Programa Fixação Biologica de Nitrogenio, Rio de Janeiro, Brazil.

I. INTRODUCTION

With the development of the acetylene reduction technique,[7,21] research related to the fixation of atmospheric nitrogen in grasses became more intense. This was due to the fact that although the existence of a symbiotic association plant-N$_2$-fixing bacteria between the cultivar "batatais" *Paspalum notatum* and *Azotobacter paspali*[8] and between *Azospirillum* spp. and some forage and grain grasses[5,11] had been proved, it was only by the use of the acetylene reduction technique that it was shown that these bacteria fixed atmospheric nitrogen.

Used at first to measure N$_2$-fixation in nodules of Leguminosae,[14] where ethylene appeared soon after addition of acetylene, the method had to be adapted to evaluate nitrogenase activity in *Gramineae* using the excised root system.[1,10] This adaptation became necessary because nitrogenase in this system presented a lag of from 8 to 18 hr before acetylene reduction began.[9,20] However, no satisfactory results were obtained from research trying to eliminate this lag by preincubation under reduced pO$_2$[9] or by addition of carbon sources, *Azospirillum* spp., NH$_4^+$ and NO$_2^-$.[1,2] Several hypotheses have been suggested to explain this lag:

1.   Oxygen inhibition of nitrogenase activity
2.   Disturbance of root N metabolism with possible accumulation of NO$_2$ or NH$_4$[6]
3.   Lack of CO$_2$ in the gaseous phase, since CO$_2$ increases nitrogenase activity in the roots[12,17]
4.   Multiplication of bacteria during preincubation due to release of organic acids resulting from fermentation, as suggested by many authors[4,6,19]

None of these hypotheses has been confirmed as being the only reason for the lag.[13] However, great concern has been expressed about the validity of the results obtained with the excised root method[3,19,20] due to errors that can result from preincubation under low pO$_2$.

The present work shows mainly the interference of NO$_2^-$ and NO$_3^-$ at the beginning of acetylene reduction in excised maize roots, with and without *Azospirillum lipoferum* inoculation. Also presented are results relating to the effect of NH$_4^+$ and glutamate in N$_2$-fixation, under the same conditions. The objective of this work is to contribute to a better understanding of the lag (latent period) without giving, however, any definite explanation.

II. MATERIAL AND METHODS

A laboratory experiment was carried out with the following treatments: (1) control, (2) NO$_3^-$, (3) NO$_2^-$, (4) NH$_4^+$, and (5) glutamate, with four replications. Maize plants, cv. Piranão, at the maturing stage, were used. The roots were collected in deionized water and taken to the laboratory where 1 g fresh weight was cut into approximately 1 cm pieces and placed in 25 mℓ flasks containing deionized water. After changing the water for N$_2$, 7 mℓ of the following solutions were added: KNO$_3$ (1.0 mM), NaNO$_2$ (0.5 mM), NH$_4$Cl (0.5 mM), and glutamic acid (0.5 mM); the control treatment received only distilled water. *Azospirillum lipoferum* nir$^+$ (denitryfying) isolated from maize roots sterilized in chloramine-T (1%) for 60 min were used in the inoculated treatment. One milliliter was employed of a bacterial suspension containing 10^7 cells/mℓ grown in liquid NF6 medium plus NH$_4$Cl concentration, and therefore without nitrogenase. It is known, however, that nitrogenase requires 3 hr to synthesize when cells are grown in NH$_4$ medium. Following injection of (10% v/v) acetylene, the oxygen level was kept close to 0.2%.

A. Determination of NO$_3$

Aliquots of 0.2 ml were taken from the flasks containing roots in solution both from the control and NO$_3$ treatments, with and without inoculation, and placed in test tubes containing a solute of 0.8 ml of 5% salicylic acid in concentration H$_2$SO$_4$. After 20 min, 9 ml of NaOH (4N) were added, and the characteristic color denoting the presence of nitrate observed. Optical density at 410 nm was then determined.

B. Determination of NO$_2$

Aliquots of 0.2 ml were taken from the flasks containing roots in solution from the control, NO$_3$, NO$_2$ treatments, with and without inoculation, and placed in flasks containing 2 ml of the 1:1 solution (v/v) of 0.02% n$_1$-naphthylethylene diamine dichlorate and 1% sulphanilamide in 1.5 M HCl, the volume being made up to 4 ml with distilled water. Absorbance at 540 nm was determined after 15 min.

C. Determination of Nitrogenase Activity

Nitrogenase activity was determined in the same flasks for all treatments by gas chromatography of the C$_2$H$_4$ resulting from C$_2$H$_2$ reduction. A gas chromatograph (Perkin Elmer F.11) equipped with 2 m × 3mm Porapak N column at 110°C was used.

D. Determination of N$_2$O

Samples were taken from two replications of each treatment after 23 hr incubation under C$_2$H$_2$ and accumulation of N$_2$O observed. A gas chromatograph (Model 7500-Portable) was used to determine the thermal conductivity, with a column of Porapak Q and using helium as the carrier-gas.

III. RESULTS

Results shown in Figure 1 indicate a decrease in the NO$_3$ level in the flasks containing roots in solution, after 3 hr, when inoculated with *Azospirillum lipoferum*, while in the control flasks there is an initial increase and a decrease after 10 hr incubation. This increase in NO$_3$ occurs possibly due to release of NO$_3$ by the roots since there is a reasonable amount of NO$_3$ in the roots (Figure 2). Figure 2 shows the disappearance of the NO$_3$ existing in the roots with and without inoculation. We may note that this disappearance is more a function of the roots' ability to release the NO$_3$ than of the bacteria to reduce it. It is also noted that the process of disappearance is the same for both inoculated and noninoculated treatments although there is a tendency, which is nonsignificant, to show during the first 6 hr smaller values of NO$_3$ in the inoculated roots.

It can be seen in Figure 1 that the disappearance of NO$_3$ is followed by accumulation of NO$_2$ in the flasks with a maximum peak after 6 hr, decreasing almost to nil after 12 hr (Figure 3). When the flasks containing NO$_3$ were not inoculated, accumulation of NO$_2$ started only after 12 hr, reaching the highest value at 18 hr. From then on the NO$_2$ content started to decrease (Figure 4).

Observing the effect of inoculation in NO$_2$ reduction, it was noted that after 6 hr incubation there was total disappearance of NO$_2$ from the inoculated flasks, while the NO$_3$ remained constant for 24 hr in the noninoculated flasks (Figures 3 and 4). On the other hand, it was noted that there was no accumulation of NO$_2$ when the roots were inoculated with *A. lipoferum* (Figure 3) indicating that the NO$_2$ produced was quickly reduced by the bacteria. However, when measuring NO$_2$ in the noninoculated roots, a small accumulation of NO$_2$ was noted, probably due to a smaller number of bacteria capable of reducing NO$_2$.

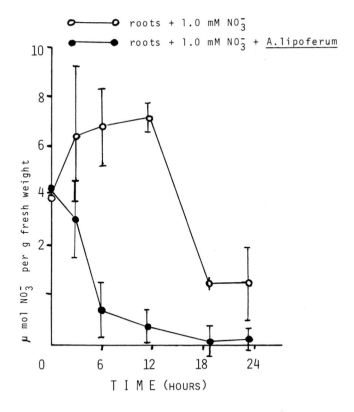

FIGURE 1.   NO₃ reduction in flasks containing maize roots with and without *Azospirillum lipoferum* inoculation. Initial NO₃ concentration in the solution was 1.0 m*M*(KNO₃).

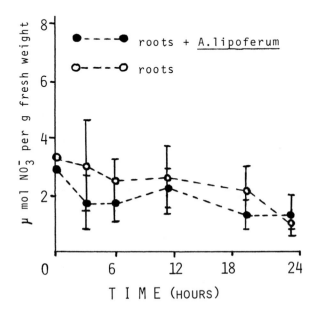

FIGURE 2.   NO₃ reduction in maize roots with and without *Azospirillum lipoferum* inoculation.

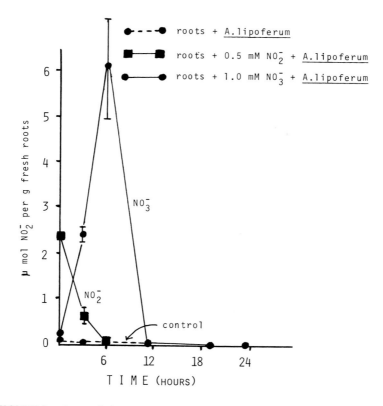

FIGURE 3. Accumulation and disappearance of NO_2^- in flasks containing maize roots inoculated with *Azospirillum lipoferum*. Initial concentration of NO_2^- and NO_3^- in the solution was 0.5 mM(NaNO₂) and 1.0 mM(KNO₃), respectively.

Figure 5 shows the effect of NO_3^-, NO_2^-, NH_4^+ and glutamate at the beginning of acetylene reduction, only in the inoculated flasks, as the noninoculated roots did not show C_2H_2 reduction. The beginning of C_2H_2 reduction was delayed in NH_4^+ and glutamate, NH_4^+ having a greater inhibiting effect than glutamate in relation to the amount of ethylene produced. No difference in the beginning of acetylene reduction was noted when comparing NO_3^- and control treatments, both showing a higher level than NH_4^+ and glutamate treatments. Nitrogenase activity in flasks with NO_3^- began after 12 hr, exactly when all NO_3^- had been reduced to NO_2^- and the NO_2^- had disappeared (Figures 1 and 3). This suggests an inhibiting effect of NO_2^- on the nitrogenase activity as suggested by Magalhães et al.[15] More clear effects are noted at the beginning of C_2H_2 reduction when NO_2^- is added (Figure 3). It was verified that the initiation of C_2H_2 reduction coincided with the disappearance of NO_2^- from the flasks after 6 hr incubation, indicating that it is necessary that all NO_2^- formed is reduced before nitrogenase activity begins.

To check on the fate of the NO_2^-, production of N_2O under C_2H_2 was determined as a measure of denitrification. Table 1 shows the accumulation of N_2O produced during 23 hr of incubation under C_2H_2. A larger accumulation of N_2O can be noted in the inoculated roots than in the noninoculated, confirming considerable denitrification by *A. lipoferum* nir⁺ (positive nitrate reductase). A larger accumulation of N_2O occurred in treatments with NO_3^-. However, treatments with NO_2^- and NH_4^+ also showed a significant increase in accumulation of N_2O when compared with the control and the glutamate treatments, but only with inoculation. There was no difference between

Table 1

N₂O ACCUMULATION IN EXCISED
MAIZE ROOTS (μmol N₂O/g FRESH
WEIGHT); 23 HR IN C₂H₂ —
AVERAGE OF TWO REPLICATIONS

Treatments	Inoculated[a]	Noninoculated
Control	1.06	0.55
0.5 mM glutamate	1.13	0.28
1.0 mM NO₃⁻	3.76	1.66
0.5 mM NO₂⁻	2.32	0.33
0.5 mM NH₄⁺	2.54	0.33

[a] *Azospirillum lipoferum* isolated from maize roots
sterilized for 60 min in 1% chloramine-T.

the other hand, the inhibiting effect of NH₄⁺ on the lag was confirmed (Figure 5) as shown by Baldani et al.[2] The smallest lag found was with the addition of NO₂⁻, contrary to data obtained by van Berkum[22] with *Azospirillum* spp. in culture media. The fact that the most quick and complete exhaustion of NO₂⁻ in flasks with roots in solution occurred when NO₂⁻ was added could explain this result. The time necessary for reduction of NO₃⁻ to NO₂⁻ can explain the smaller rate of nitrogenase activity found with NO₃⁻ addition.

The masking effect of nitrate reductase was amply shown with the inoculation of *A. lipoferum* (Table 1). The largest accumulation of N₂O with the addition of NO₃⁻ extended to *A. lipoferum* the results previously obtained with *A. brasilense*.[18] However, NH₄⁺ stimulated the production of N₂O in the inoculated flasks. This indicates that *A. lipoferum* nir⁺, possibly associated with nitrifying bacteria, can stimulate NH₄⁺ dentrification.

REFERENCES

1. **Abrantes, G. T. V., Day, J. M., and Döbereiner, J.,** Methods for the study of nitrogenase activity in field grown grass, *Bull. Int. Infor. Bio. Sol Lyon,* 21, 1, 1975.
2. **Baldani, J. I., Pereira, P. A. A., Neyra, C. A., and Döbereiner, J.,** The initiation of acetylene reduction in isolated roots of maize. Effect of carbon, oxygen and mineral nitrogen sources, in *Limitations and Potentials of Biological Nitrogen Fixation in the Tropics,* Basic Life Science, Vol. 10, Döbereiner, J., Burris, R., and Hollander, A., Eds., Plenum Press, New York, 1978, 356.
3. **Barber, L. E., Tjepkema, J. D., and Evans, H. J.,** Nitrogen Fixation in the Roots Environment of Some Grasses and Other Plants in Oregon, Int. Symp. Environmental Role Nitrogen-Fixing Blue-Grass Algae and Asymbiotic Bacteria, Uppsala, Sweden, 1976.
4. **Barber, L. E. and Evans, H. J.,** Characteristics of nitrogen fixing bacterial strain from the roots of *Digitaria sanguinalis, Can. J. Microbiol.,* 22, 254, 1976.
5. **von Bülow, J. F. W. and Döbereiner, J.,** Potential for nitrogen fixation in maize genotypes in Brasil, *Proc. Nat. Acad. Sci. U.S.A.,* 72, 2389, 1975.
6. **Day, J. M., van Berkum, P., and Witty, P. F.,** Associative Symbioses: Their Contribution to the Nitrogen Cycle and a Critical Analysis of the Methodology Used in Their Study, Int. Symp. Limitations Potentials Biological Nitrogen Fixation in Tropics, Brasília, 1977.
7. **Dillworth, M. J.,** Acetylene reduction by nitrogen fixing preparations from *Clostridium pasteurianum, Biochem. Biophys. Acta,* 127, 285, 1966.

8. Döbereiner, J., Further research on *Azotobacter paspali* and its variety specificity occurrence in the rhizosphere of *Paspalum notatum, Zentralbl. Bacteriol. Parasitenk. Infectionskr. Hyg. Abt.,* 124, 224, 1970.

9. Döbereiner, J., Day, J. M., and Dart, P. J., Nitrogenase activity and oxygen sensibility of the *Paspalum notatum Azotobacter paspali* association, *J. Gen. Microbiol.,* 71, 103, 1972.

10. Döbereiner, J., Day, J. M., and Dart, P. J., Fixação de nitrogênio na rizosfera de *Paspalum notatum* e de cana-de-açúcar, *Pesqui. Agropec. Bras.,* 8, 153, 1973.

11. Döbereiner, J. and Day, J. M., Associative Symbiosis in Tropical Grass: Characterization of Micro-organisms and Dinitrogen Fixing Sites, presented at International Symposium on N_2-Fixation, Interdisciplinary Discussions, June 3—7, 1974, Washington State University, Pullman, 1974.

12. Döbereiner, J., Physiological Aspects of N_2-Fixation in Grass-Bacteria Associations, 2nd Int. Symp. N_2-fixation, Salamanca, 1976.

13. Döbereiner, J., Nitrogen fixation in grass — bacteria associations in the Tropics, in *Isotopes in Biological Dinitrogen Fixation,* International Atomic Energy Agency, Vienna, 1978, 51.

14. Koch, B. and Evans, H. J., Reduction of acetylene to ethylene by soybean root nodules, *Plant Physiol.,* 41, 1748, 1966.

15. Magalhães, L. M. S., Neyra, C. A., and Döbereiner, J., Nitrate and nitrate reductase negative mutants of N_2-fixing *Azospirillum* spp., *Arch. Microbiol.,* 117, 247, 1978.

16. Neyra, C. A. and van Berkum, P., Nitrogenase activity in isolated sorghum roots. Effect of bicarbonate and inorganic nitrogen, *Plant Physiol.,* 57, S-533, 1976.

17. Neyra, C. A. and van Berkum, P., Nitrate reduction and nitrogenase activity in *Spirillum lipoferum, Can. J. Microbiol.,* 23, 306, 1977.

18. Neyra, C. A., Döbereiner, J., Lalande, R., and Knowles, R., Dentrification by N_2-fixing *Spirillum lipoferum, Can. J. Microbiol.,* 23, 300, 1977.

19. Okon, Y., Albrecht, S. L., and Burris, R. H., Methods for growing *Spirillum lipoferum* and for counting it in pure culture and in association with plants, *Appl. Environ. Microbiol.,* 33, 85, 1977.

20. Rinaudo, G., Balandreau, J., and Dommergues, Y., Algal and bacterial non-symbiotic nitrogen fixation in paddy soil, in *Nitrogen Fixation in Natural and Agricultural Habits, Plant and Soil Special,* 1971, 471.

21. Schollhorn, R. and Burris, R. H., Acetylene as a competitive inhibition of N_2-fixation, *Proc. Nat. Acad. Sci. U.S.A.,* 58, 213, 1967.

22. van Berkum, P. B., Nitrogenase Activity with Tropical Grass Roots and Some Effects of Combined Nitrogen on *Spirillum lipoferum* Beijerinck, Ph.D. thesis, London University, England, 1978.

* Okinawa Branch, Tropical Agricultural Research Center, Ishigaki, Okinawa, Japan.

I. INTRODUCTION

Enhancing nitrogen-fixing activity in grass-bacteria associations would be of economic importance especially in the case of forage grasses, cover crops and green manures, because the application of chemical nitrogen fertilizer to these crops may put the cost out of balance.

It was reported that several kinds of forage grasses had associations with nitrogen-fixing bacteria, showing considerable nitrogenase activity in their rhizospheres. The dominant nitrogen-fixer in each association has also been investigated. Considering the environmental conditions under which promising nitrogen-fixers can exist, efficient grass-bacteria associations might be found not only in the American continents but in other regions such as tropical Asia.

Pennisetum purpureum (Napiergrass) is one of the most productive forage crops, with the efficient C_4 photosynthesis, and reported to have considerable nitrogenase activity in its root.[2] The present paper reports the factors regulating nitrogen fixation in *Pennisetum* root, and the distribution and population of nitrogen-fixing bacteria in its rhizosphere.

II. MATERIALS AND METHODS

The experiments described below were carried out at Okinawa Branch, TARC. It is located near the Tropic of Cancer, and the yearly mean temperature at that site is 23.6°C with a yearly mean precipitation of 2090 mm. The yearly minimum soil temperature (20 cm below the surface) is about 15°C.

A. Plant Materials

Pennisetum purpureum was planted in our experimental field where the soil pH was 4.7 to 5.0. Fertilizers of 20 kg N, 80 kg P and 80 kg K/ha were applied together with 1 kg of diluted sodium molybdate. Between 6 to 10 months after planting, the root systems were separated from the plant tops and washed in sterilized water to free them from soil.

B. Assay for Nitrogenase Activity

The method used for measuring nitrogenase activity was similar to that reported by Abrantes et al.[1] with slight modifications. Before the assay, root samples were washed twice with sterilized water and placed in vessels. These look like a small dessicator with a volume of 270 mℓ, with a stopcock and a serum stopper on its cover. The vessels were evacuated three times to 80 mm Hg and refilled with the desired gas mixtures.

In order to supply effective carbon sources for acetylene reduction by the root, 5 mℓ of various carbon source solutions were applied to ca. 1 g (dry weight) of root at the beginning of preincubation. Concentration of carbon source solutions was 50 mM, and the pH of organic acid solutions was adjusted to 6.0 with NaOH.

Acetylene reduction was carried out with 10% acetylene for scheduled periods at room temperature (25 to 27°C). Ethylene produced was measured on a 1 mℓ gas sample collected through a serum stopper with a Yanako G-80 hydrogen flame ionizing gas chromatograph fitted with a 2 m × 3 mm Porapak R column using nitrogen as a carrier gas at 50°C.

C. Estimation of the Distribution and Population of Nitrogen Fixing Bacteria in the Rhizosphere

Most probable number (MPN) technique was used for determining the distribution

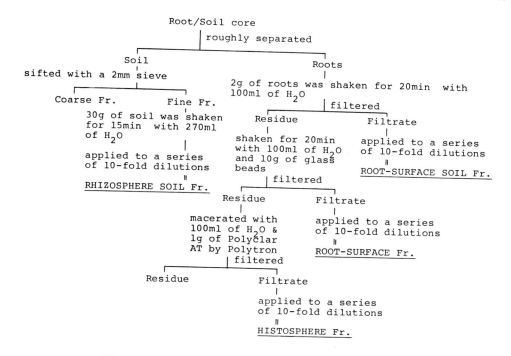

FIGURE 1. Procedure for fractionating Root/Soil core samples.

and population of nitrogen-fixing bacteria in the rhizosphere of *Pennisetum purpureum*. Figure 1 shows the procedure used for fractionating a root soil core sample into four fractions. Polyclar-AT was used for absorbing polyphenols exuded from roots in maceration. Series of tenfold dilution of each fraction were prepared and 0.2 mℓ of aliquot was inoculated into a 7 mℓ vial with 3 mℓ of semisolid medium. After incubation for 48 hr at 30°C, acetylene reduction was assayed for 1 hr at 30°C, adding 10% acetylene. Vials which produced over 1 n mol ethylene per hour were considered positive and referred to MPN table.

The glucose medium used for counting was 0.5% glucose, 0.05% K_2HPO_4, 0.02% $CaCl_2 \cdot 2H_2O$, 0.02% $MgSO_4 \cdot 7 H_2O$, 0.05% $FeSO_4 \cdot 7H_2O$, 0.001% $MnSO_4 \cdot 4H_2O$, 0.0005% $Na_2MoO_4 \cdot 2H_2O$, 0.1% $CaCO_3$. In Na-malate medium, glucose was replaced by Na-malate and $CaCO_3$ with 0.1% NaCl. Concentration of calcium chloride was decreased to 0.002%. Additional contents of both media were as follows, 50 mg of yeast extract, 2 mℓ of 0.5% BTB and 1.75 g of soluble agar per liter.

III. RESULTS AND DISCUSSION

A. Effect of Oxygen Tension on Nitrogenase Activity in the Root

Figure 2 shows the nitrogenase activity in *Pennisetum* root as affected by oxygen tension. Figure 2a shows the time-course characteristics without preincubation, while Figure 2b shows results after preincubation for 17 hr. An oxygen concentration of 2% maximized the acetylene reduction by the root, and the preincubation doubled the rate. As expected, 20% oxygen concentration strongly reduced the rate to about one fifth. This result initiated further experiments under an oxygen concentration of 2% after preincubation for 17 hr.

B. Effect of Root Age on the Nitrogenase Activity

Nitrogenase activity changed remarkably with the age of the root used (Table 1).

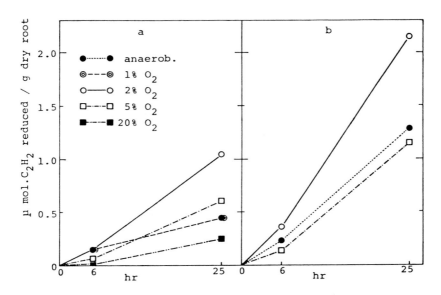

FIGURE 2. Time-courses of acetylene reduction by *Pennisetum* root incubated at various oxygen concentrations; (a) without preincubation, (b) acetylene was introduced after preincubated for 17 hr at each oxygen concentration.

Table 1
**EFFECT OF ROOT AGE ON THE
NITROGENASE ACTIVITY IN THE
ROOT**

Positions of roots developed	nmol.C_2H_2reduced/hr·g dry root
Original cutting and primary stalks	28[a] ± 3[b]
Secondary stalks	19 ± 3
Tertiary stalks	1 ± 0.3
Fourth stalks and tertiary stalks without tillering	1 ± 0.3

[a]   Means of three replicates.
[b]   Standard error.

The younger the root, the lower the activity. Döbereiner et al.[3] reported that mature thick roots were always most active in both C_3 and C_4 grasses examined. This coincides well with the above findings. These results suggest that having a uniform root age may minimize the fluctuation of acetylene reduction by the root. A similar result might be obtained if the root samples were separated according to their color gradient, because the surface color is considered to reflect the maturity.[5]

C. Enhancement Effect of Exogenous Carbon Sources on Nitrogenase Activity in the Root

In order to determine the effective carbon sources for acetylene reduction by *Pennisetum* roots, various carbon source solutions were applied to the root at the beginning of preincubation. As shown in Table 2, sugars such as glucose and mannitol markedly enhanced the capacity for reducing acetylene in the root, while malic acid

Table 2
ENHANCEMENT EFFECT OF EXOGENOUS CARBON SOURCES ON ACETYLENE REDUCTION BY THE ROOT

Carbon source	nmol.C_2H_2 reduced/ hr·g dry root
H_2O as control	$40^a \pm 10^b$
Glucose	105 ± 3
Mannitol	167 ± 17
Malic acid	52 ± 8
Succinic acid	95 ± 20
Pyruvic acid	79 ± 16

[a] Means of three replicates.
[b] Standard error.

Table 3
DISTRIBUTION AND POPULATION OF NITROGEN-FIXING BACTERIA IN THE RHIZOSPHERE OF *PENNISETUM PURPUREUM*

Fraction	Carbon source of medium	
	Glucose	Na-malate
Rhizosphere soil	5×10^{2a}	8×10^2
Root-surface soil	43×10^{3b}	2×10^3
Root-surface	200×10^{3b}	5×10^3
Histosphere	$23,000 \times 10^{3b}$	12×10^3

[a] Per gram of soil.
[b] Per gram of fresh root in lower three fractions.

slightly enhanced. This result is a sharp contrast to the result obtained by van Berkum et al.[6] They reported that the addition of malate and bicarbonate doubled the nitrogen fixation by isolated sorghum roots while glucose had no effect.

D. Distribution of Nitrogen-Fixing Bacteria in the Rhizosphere

According to the results shown in Table 3, the population of nitrogen-fixers increased in the histosphere fraction compared with others. The glucose medium recovered more than malate medium in the lower three fractions, resulting in a remarkable difference in the histosphere fraction.

This may reflect the existence of a considerable number of nitrogen-fixing bacteria which cannot use malic acid as a carbon source. This difference in microflora might result in the different enhancement effect of exogenous carbon sources on nitrogen fixation in the two experiments mentioned above.

In conclusion, it is suggested that the content of sugars in the root is one of the main factors regulating nitrogen fixation by *Pennisetum* root. Since differences in environmental conditions might result in dissimilarity of microflora in the grass rhizosphere, effective associations between grass and bacteria would vary in different regions.

REFERENCES

1. **Abrantes, G. T. V., Day, J. M., and Döbereiner, J.,** *Bull. Int. Infor. Biol. Sol Lyon*, 21, 1, 1975.
2. **Day, J. M., Neves, M. C. P., and Döbereiner, J.,** *Soil Biol. Biochem.*, 7, 107, 1975.
3. **Döbereiner, J., Day, J. M., and von Bülow, J. F. W.,** Proc. 2nd Int. Winter Wheat Conf., Zagreb, Yugoslavia, 1975, 221.
4. **Döbereiner, J., in** *Limitations and Potentials for Biological Nitrogen Fixation in the Tropics,* Döbereiner, J., Burris, R., and Hollander, A., Eds., Plenum Press, New York, 1978, 13.
5. **Fujii, Y.,** (Japanese with English summary), *Agric. Bull. Saga Univ.*, 12, 1, 1961.
6. **van Berkum, P. and Neyra, C. A.,** *Plant Physiol.*, 57, s-533, 1976.

Chapter 9

SPECIFICITY OF INFECTION BY *AZOSPIRILLUM* SPP. IN PLANTS WITH C₄ PHOTOSYNTHETIC PATHWAY*

R. E. M. Rocha, J. I. Baldani, and J. Döbereiner**

TABLE OF CONTENTS

*   Translator: Diva Athié.
** Programa Fixação Biologica de Nitrogenio, Rio de Janeiro, Brazil.

I. INTRODUCTION

Host specificity in *Gramineae* was initially demonstrated through the symbiotic association of a *Paspalum notatum* ecotype with *Azotobacter paspali*.[3] This association seems to be more restrictive than symbiosis with leguminous plants, as it is limited to some ecotypes of one species.[4]

Great interest in this association arose after biological N_2-fixation in grain and forage grasses was shown to be caused by *Azospirillum* spp.[2,6] Patriquin and Döbereiner[7] demonstrated infection by *Azospirillum* spp. of cortical and stelar tissue of maize roots, confirming the close plant/bacteria interaction.

Recent studies[1] showed host specificity in maize, wheat, and rice, in root infection by *Azospirillum* spp. While wheat and rice are more susceptible to infection by *Azospirillum brasilense* nir⁻, maize roots seem to be mainly infected by *Azospirillum lipoferum*.

II. EXPERIMENTAL

With the objective of widening and supporting the results already obtained,[1] host specificity was studied in a group of plants, comprising eight forage grasses and one *Cyperacea* having the C_4 photosynthetic pathway. Soil and root samples were collected at five different places where there was a predominance of the majority of the *Gramineae* studied. The soil taken from the various places was identified as Planasol, Agrostology and Ecology series, and Red-Yellow Podzolic, Itaguai series.

Azospirillum spp. were isolated from soil and roots with and without sterilization in chloramine-T, using the method described by Döbereiner and Baldani.[5] Identification of the species was made in accordance with Tarrand et al.[8]

Occurrence of *Azospirillum* spp. was noted in 100% of the soil samples and 80% of the sterilized root samples. From the plants studied 100% of the soil isolates were identified as *A. brasilense*, while in nonsterilized roots the percent of *A. brasilense* and *A. lipoferum* varied from plant to plant.

Although *A. brasilense* was predominant in the soil, the isolates from sterilized roots were almost exclusively *A. lipoferum*. Of the nine C_4 species studied, eight (including the *Cyperacea*) showed more than 80% *A. lipoferum*, thus indicating specific selection of strains by the roots of C_4 plants. The only exception occurred in sugar cane where *A. brasilense* was more frequently found (Table 1).

The results obtained indicated that the C_4 forage grasses and the *Cyperacea* studied seem to be preferentially infected by *A. lipoferum* in contrast to C_3 which are mostly infected by *A. brasilense* nir⁻.[9] It is supposed, therefore, that the physiology of the C_4 plants make them susceptible to infection by *A. lipoferum*, but not by *A. brasilense*. There is no explanation why this occurs.

Table 1
HOST SPECIFICITY IN *AZOSPIRILLUM* SPP. AND C₄ *GRAMINEAE* ASSOCIATIONS

Species	Sample[a]	Sterilization time (min)	No. of isolates	% isolates[b] A. lipoferum	A. brasilense
	Soil	0	5	0	100
Brachiaria spp.	Roots	0	19	74	26
	Roots	1	12	100	0
	Soil	0	5	0	100
Panicum maximum	Roots	0	12	67	33
	Roots	0.5	5	100	0
Cynodon dactylon	Roots	0	9	78	22
	Roots	0.5	7	100	0
Pennisetum purpureum	Roots	0	14	86	14
	Roots	0.5	3	100	0
Hordeum altimissima	Roots	0	16	81.3	18.7
Digitaria decumbens	Roots	0	2	100	0
Cyperus rotundus	Soil	0	5	0	100
	Roots	0.25	5	100	0
	Soil	0	5	0	100
Sorghum vulgare	Roots	0	8	25	75
	Roots	15	14	100	0
	Soil	0	5	0	100
Saccharum spp.	Roots	0	4	75	25
	Roots	1	24	0	100

[a] Soil samples collected at different places and with different soils.
[b] Percent calculated on the number of isolates in each treatment.

REFERENCES

1. **Baldani, V. L. D. and Döbereiner, J.,** Host plant specificity in the infection of cereals with *Azospirillum* spp., *Soil Biol. Biochem.*, 1979, in press.
2. **von Bülow, J. F. W. and Döbereiner, J.,** Potential for nitrogen fixation in maize genotypes in Brasil, *Proc. Nat. Acad. Sci. U.S.A.,* 72, 2389, 1975.
3. **Döbereiner, J.,** *Azotobacter paspali* sp. n., uma bactéria fixadora de nitrogênio na rizosfera de *Paspalum notatus, Pesqui. Agropecu. Bras.,* 1, 357, 1966.
4. **Döbereiner, J.,** Further research on *Azotobacter paspali* and its variety specific occurrence in the rhizosphere of *Paspalum notatus, Zentralbl. Bacteriol. Parasitenk. Infectionskr. Hyg. Abt. 2,* 124, 224, 1970.
5. **Döbereiner, J. and Baldani, V. L. D.,** Prospects for inoculation with *Azospirillum* spp., in *Associative N₂-Fixation,* Vose, Peter B. and Ruschel, A. P., Eds., CRC Press, Boca Raton, Fla., 1981.
6. **Döbereiner, J. and Day, J. M.,** Associação de bactérias fixadoras de nitrogenio com raizes de gramineas, Porto Alegre, Reunião Latinoamericana do Trigo, 1974.
7. **Patriquin, D. G. and Döbereiner, J.,** Bacteria in the endorhizosphere of maize in Brasil, *Can. J. Microbiol.,* 24, 734, 1978.
8. **Tarrand, J. J., Krieg, N. R., and Döbereiner, J.,** A taxonomic study of the *Spirillum lipoferum* group, with description of a new genus *Azospirillum* gen. nov. and *Azospirillum brasilense* sp. nov., *Can. J. Microbiol.,* 24, 967, 1978.
9. **Baldani, V. L. D. and Döbereiner, J.,** unpublished.

Chapter 10

FIXATION OF NITROGEN BY *HYPARRHENIA RUFA* NEES, IN SITU*

K. M. Tshitenge**

TABLE OF CONTENTS

*   Translator: P. B. Vose.
** Centre Régional d'Etudes Nucléaires de Kinshasa, Kinshasa, Zaire.

I. INTRODUCTION

The interaction which exists between grasses and certain soil bacteria and which results in the reduction of molecular nitrogen is very complex. The amount depends on the genotype of the plant[5,6,7] and the efficiency of light conversion.[1,3] It equally depends on the bacteria. In certain cases, a single bacteria forms an association with the grass which is especially capable of fixing atmospheric nitrogen. This is the case notably with *Azotobacter paspali* which, associated with *Paspalum notatum,* is able to fix up to 90 kg N_2/ha/year.[5] In other cases the fixation seems to result from a synergistic effect of several bacteria situated in the rhizosphere of the plants, e.g., sugar cane.[7]

Nitrogen fixation in "associative symbiosis" of grasses equally depends on climatic conditions, such as air and soil temperature, and the nature and pH of the soil. The present work reports the results obtained in a study of nitrogen fixation by *Hyparrhenia rufa* in situ.

II. MATERIALS AND METHODS

A cylindrical, transparent, and leakproof polyethylene bag, 49 × 20 cm, was placed and made gas-tight by means of Apiezon sealing compound, on a concrete pot 49 × 39 cm. The pot had three fourths of its volume filled with soil and contained five vigorous plants of *H. rufa*. These had been established for 4 years and had received no fertilizer for 3 years. The gas-tightness of the system was then checked with acetylene and gas chromatography. Acetylene (10% of total volume) was produced, following the reaction of 10 g calcium carbide with 2.8 mℓ H_2O. The reduction of acetylene was followed in a time course, by means of gas chromatography using a Beckman® GC65 gas chromatograph.

III. RESULTS AND DISCUSSION

Figure 1 shows the development of nitrogenase activity. It can be seen that during the course of a day the curve shows a maximum corresponding approximately to midday; from then the apparent fixation of nitrogen starts to decline, attaining only low values at the end of the day. At the start of the following day nitrogen fixation recommences anew and attains a maximum about midday; nevertheless this peak is not as high as the first.

In a second experiment carried out with the same material a week after, having left the plants to recover, other measurements were taken during the night. Although the values of ethylene produced were lower than those reported for the first experiment, the curve presented the same shape as the first, with a maximum in the middle of the day. Afterwards the fixation declined markedly, to stabilize during the course of the night. The next day, nitrogen fixation recommenced and showed a slight peak about midday.

The lower height of the second peak (Figure 1, A and B) could be due to the fact that the acetylene-ethylene gas mixture is toxic to the plants. These results agree with those that have been reported by Balandreau and Villemin[3] and Balandreau[1] for cultivars of the same species. The low values obtained during the second experiment could be explained by the toxicity of the acetylene and/or by the temperature inside the polyethylene bag. It was indeed observed that after the first experiment many leaves were withered or necrotic.

The quantities of nitrogen fixed, extrapolated from the values of ethylene produced, varied between 25 kg N_2/ha/year (night) and 40 kg N_2/ha/year (day). Integrating the

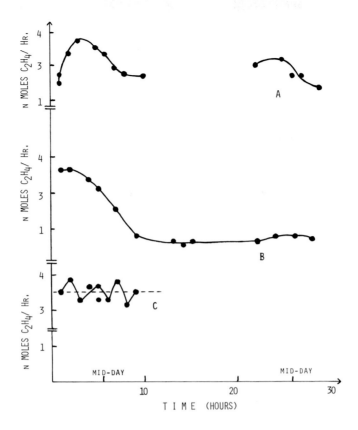

FIGURE 1.   Development of nitrogenase activity.

area derived from the curve, a mean value equal to 49 kg N_2/ha/year is obtained. This value is close to those which have been reported by Balandreau and Villemin[3] and Balandreau.[1] It is nevertheless lower than that reported by Döbereiner et al.[5] for the *A. paspali-P. notatum* association which is equal to 90 kg N_2/ha/year. However, all this type of data is subject to both the error of extrapolation and the theoretical conversion of ethylene production to N_2-fixation.

With the objective of determining the influence of the aerial parts and consequently photosynthesis, on the fixation of nitrogen, the activity of nitrogenase was followed in *H. rufa* from which all the leaves had been removed. The study was carried out on the same pot and the same plants, except that two of the plants had been removed for another study. Figure 1, C shows the result obtained. It can be seen that the graph of nitrogenase activity remains more or less constant during the whole day. This fact suggests that fixation of nitrogen in *H. rufa* depends on the amount of photosynthesis. Thus the nitrogen fixation in *H. rufa* without leaves remains constant the whole time, despite the intact system showing a peak around midday.

One can equally note that the values for quantity of nitrogen fixed are greater than those shown in the graph of Figure 1, B; in this case the amount of N_2 reduced is equal to 45 kg N_2/ha/year. This value is close to that obtained for the entire system. However it is known that the amount of carbohydrate exudates, the pH resulting, the temperature of the soil, and the breakdown of the plant root system can modify the equilibrium of the rhizosphere unfavorably for nitrogenase activity.

ACKNOWLEDGMENTS

Dr. Alaides Ruschel is thanked for guidance during this work, together with all the nitrogen fixation group at CENA, and especially Luiz A. Gracioli for discussion of the results. This work was carried out during the tenure of an International Atomic Energy Agency Type II Fellowship, and thanks are due to IAEA and the Comissão Nacional de Energia Nuclear of the Government of Brasil.

REFERENCES

1. **Balandreau, P. J.,** Limiting Factors in Grass N$_2$-fixation, in Int. Symp. on the Limitations and Potentials of Biological Nitrogen Fixation in the Tropics — Summary of Papers, Universidade de Brasília, July 18 to 22, 1977.
2. **Balandreau, P. J., Millier, C. R., and Dommergues, Y. R.,** Mesure de l'activité nitrogenasique des microorganismes fixateurs libres d'azote de la rhizosphère de quelques graminées, *Rev. Biol. Sol,* 12(1), 259, 1974.
3. **Balandreau, P. J. and Villemin, G.,** Fixation biologique de l'azote moleculaire en savane de Lamto (Basse Côte d'Ivoire). Resultats préliminaires, *Rev. Ecol. Biol. Sol,* 10, 25, 1973.
4. **Day, J. M., Neves, M. C. P., and Döbereiner, J.,** Nitrogenase activity on the roots of tropical forage grasses, *Soil Biol. Biochem.,* 7, 107, 1975.
5. **Döbereiner, J., Day, J. M., and Dart, P. J.,** Nitrogenase activity in the rhizosphere of sugar cane and some other tropical grasses, *Plant Soil,* 37, 191, 1972.
6. **Larson, R. I. and Neal, J. L.,** Selective Colonization of the Roots of Wheat by N$_2$-Fixing Bacteria, Abstracts Proc. Int. Symp. Environ. Role of Nitrogen Fixing Blue-grass Algae and Assymbiotic Bacteria, Uppsala, Sweden, September 1976.
7. **Ruschel, A. P. and Vose, P. B.,** Present Situation Concerning Studies on Associative Nitrogen Fixation in Sugar Cane, *Saccharum officinarum* L., CENA-Bol. Cientifíco BC-045, Centre de Energia Nuclear na Agricultura, Piracicaba, S. P., Brasil, July 1977.

Chapter 11

THE ECONOMIC IMPORTANCE OF ASSOCIATIVE N_2-FIXATION IN SEMIARID GRASSLAND REGIONS OF ARGENTINA

Anibal H. Merzari and Dolly Carpio*

TABLE OF CONTENTS

* Center of Radiobiology, Faculty of Agronomy, University of Buenos Aires, Argentina.

I. INTRODUCTION

The problems connected with the use of new sources of energy and especially those of renewable character, are very important in our times in relation to the petroleum crisis.

Argentina, now produces 90% of the fuel necessary for its requirements and is not apparently affected at the moment, but it is useful to remember that Argentina is a country with petroleum but is not a seller of petroleum. The known reserves in our country are enough for only 12 years. We must be obliged to consider carefully this problem and to require immediate action to develop a program on the use of nonconventional sources of energy.

Using chemical fertilizers we use petroleum. Only biological nitrogen fixation is our long term solution for the large areas of land under consideration. We must therefore demonstrate the significance and the possibilities of associative dinitrogen fixation in relation to the big area of Argentina where it is possible to expand livestock, bearing in mind that to produce protein we need nitrogen.

II. DISTRIBUTION AND CHARACTERISTICS OF GRASSLAND

In Figure 1 it can be seen that the territory of Argentina possesses natural grasslands with different possibilities. These grasslands cover practically all the country.

Inside that big area is situated the semiarid region (Figure 2) and this region can be divided into two subregions: chaqueña and pampeana. The first covers an area of 20 million hectares of Argentina.[3]

The subregion of pampeana comprises the south of the province of Santiago del Estero, a little part of the northwest of the province of Santa Fe, west of the province of Córdoba, the center and the east of La Pampa, and west of the province of Buenos Aires. This subregion extends to the west in the province of San Luis on the limit with Mendoza. Here there is located natural pasture that comprises an area of two million hectares.

Referring to the subregion of chaqueña, in agreement with the opinion of Molina,[3] we may note that this is a very important reserve for Argentina. Here there is the possibility of very good agricultural development, and livestock with increased yields.

Similarly, in the subregion pampeana, according to Anderson, Del Aguilla and Bernardón[1] writing about the natural grasslands of the province of San Luis, it was noted that in that province the grassland is the most important resource. Of the area 85% is under this kind of management.

The grasses we have found in the semiarid region belong to a number of genera, as shown in Table 1.[2] It will be noted that we have determined N₂-fixation and the presence of *Azotobacter, Beijerinckia* and *Azospirillum* in a majority of the genera. Clearly the contribution of associative N₂-fixation to the nitrogen economy of this region must be considerable.

It is very important to note that many areas of the semiarid region are degraded due to excess of grazing animals and there is some data[4] to show the importance of the presence of *Gramineae* in relation to the amount of nitrogen in this soil. For instance, in the north of this region (west of the province of Formosa) it was possible to increase the total amount of nitrogen in the soil by 35% only from allowing the regeneration of the original grassland (see Table 2).

FIGURE 1.   Grasslands in Argentina.

III. ECONOMICS

At this stage attempting to put a monetary value on N_2-fixed by associative N_2-fixation in these grasslands must be largely notional. However, if we consider that this semiarid region has a grassland area of 50 million hectares, and let us assume a nitrogen fixation of only 5 kg/ha/year, we arrive at a total fixation of 250 million kg of nitrogen per year. In dollars this has a value of nearly U.S. $400 million at present prices. Although this is a hypothetical calculation it is sufficient to show the immense economic importance of associative N_2-fixation in Argentina's grasslands. Moreover it clearly indicates the importance of further studies to optimize nitrogen fixation through the most appropriate grass genera and through grassland management.

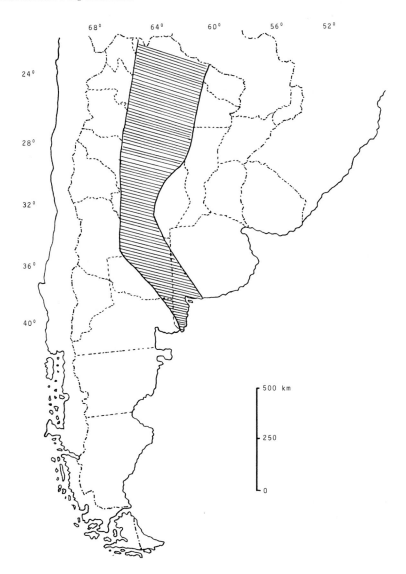

FIGURE 2.   The semiarid region of Argentina.

Table 1
GENERA OF *GRAMINEAE*
FOUND IN THE GRASSLANDS OF
THE SEMIARID REGION OF
ARGENTINA

Subregion pampeana	Subregion chaquena
Aristida	
Botriochloa	
Briza	*Pennisetum*
Bromus	*Andropogon*
Cenchrus	*Sporobulus*
Cynodon	*Gouinia*
Chloris	*Panicum*
Digitaria	*Leptochloa*
Elyonurus	*Cenchrus*
Eragrostis	*Elyonurus*
Pappophorum	*Eragrostis*
Poa	*Chloris*
Setaria	*Aristida*
Sorghastrum	
Sporobulus	
Stipa	
Trichloria	
Bouteloua	
Festuca	
Paspalum	
Eleusine	
Pennisetum	

Note: In the genera underlined we determined N_2-fixation and the presence of *Azotobacter, Beijerinckia* and *Azospirillum*[2].

Table 2
PERCENT OF ORGANIC MATTER AND
TOTAL NITROGEN IN SOILS OF ING.
JUÁREZ (PROVINCIA DE FORMOSA)

Sample	Organic matter	Total nitrogen
Grassland regenerated with natural *Gramineae*	1.84	0.11
Grassland degraded by overgrazing	0.89	0.081

From Piñeiro, A., Rev. *Agron. Noroeste Argent.,* 3 (1 and 2), 1959. With permission.

REFERENCES

1. Anderson, D. L., Del Aguila, J. A., and Bernardón, A. E., *Rev. Agropec. INTA, Ser. 2,* 7(3), 153, 1970.
2. Carpio, D. and Merzari, A. H., unpublished, 1979.
3. Molina, J. S., *Cienc. Invest.,* 29, 306, 1973.
4. Piñeiro, A., *Rev. Agron. Noroeste Argent.,* 3 (1 and 2), 1959.

I. INTRODUCTION

Fixation of nitrogen by some association with sugar cane has been suspected for some time, due to the fact that the crop frequently only responds poorly to nitrogen fertilization,[2,16,51,52] and the fact that in some areas cane is regularly cut and removed from the field yet comparatively little or even no fertilizer nitrogen is supplied.

A crop of sugar cane (100 t/ha/[1]/yr) in Brazil removes typically 132 kg of nitrogen, plus the burned leaves which contain 35 to 50 kg N/ha.[26] Despite this, even on soils with poor nutrient levels, cane crops have been grown continuously for 30 years or more without N fertilizer.[2,3] In such cases the balance of nitrogen required for growth must come from long time soil-N reserves, rainfall, and free living associative N_2-fixing microorganisms.

Association of N_2-fixing microorganisms with sugar-cane roots was demonstrated by Döbereiner[11,12] who found higher populations of *Beijerinchia* sp. in rhizosphere soil than in soil between rows. This was subsequently confirmed by acetylene reduction technique, when sugar-cane roots were found to produce C_2H_4 from C_2H_2 at a rate of up to 5 mol/g/h, indicating the presence of nitrogenase.[14] This was subsequently verified at CENA and subsequent work using ^{15}N showed[35] direct evidence for dinitrogen fixation, presumably originating from microorganisms inhabiting the root, and also confirming translocation of the fixed nitrogen to the plant tissues.

II. DIRECT EVIDENCE OF NITROGEN FIXATION

Ruschel et al.[35] showed that with sugar-cane seedlings grown in compost-soil mixture, ^{15}N was fixed by L61-41 seedlings in an experiment conducted at a $pO_2 = 0.02$ after 30 hr of incubation. Much lower, but still significant levels of N-fixation were obtained with seedlings of plant L64-43, least ^{15}N being fixed at $pO_2 = 0.16$. This is in accordance with observations of Döbereiner et al.[13] on the effect of O_2 pressure on C_2H_2 reduction in sugar-cane roots.

The high amount of ^{15}N which accumulated in the leaves of the L61-41 offspring of $pO_2 = 0.02$ after only 30 hr of incubation suggested translocation of the fixed nitrogen from the roots without its prior incorporation into microbial proteins as an intermediate step. The values of 5.55 and 0.17 nmol/g/hr of C_2H_2 reduced by sugar-cane roots at pO_2 of 0.04 and 0.20, respectively, obtained by Döbereiner et al.[13] are equivalent to 2.072 and 0.0634 µg nitrogen in 40 hr on the basis of the theoretical ratio of 3:1 mol of C_2H_2 produced to N_2-fixed. This is comparable to 1.06 µg ^{15}N/g and 21.5 µg ^{15}N/g fixed by L64-43 roots during 40 hr at $pO_2 = 0.16$ and 0.02, respectively. Higher amounts of ^{15}N were fixed by the whole plant.

Many studies on nonsymbiotic nitrogen fixation have been done with soil-plant systems under drastically-modified, rather than *in situ* conditions.[14,21,30,35,50] So, in later work[42] N_2-fixation of soil sugar cane intact systems were compared with that of the disturbed ones. No prominent differences in 15-nitrogen fixation between plants of intact system and plants with soil adhering to roots were observed. Addition of glucose increased N_2-fixation in disturbed plants incubated with [15]N_2 in an atmosphere containing 2% oxygen. This may indicate that sugar cane has a potential to increase N-fixation since it has been shown by Hartt et al.[22] that labeled sucrose arriving at the roots appeared to move out into the nutrient solution.

Except for one experiment with a disturbed system, fixed nitrogen was not found in the leaves of either intact or interrupted systems after 24 hr of experimental conditions under [15]N_2 atmosphere. Intact systems kept under simulated normal atmosphere conditions for 6 days showed considerable amounts of fixed 15-nitrogen in the leaves, presumably following transfer of fixed nitrogen from the roots.

Carbon dioxide at a concentration 200-fold higher than that of air did not affect nitrogen fixation. Increase in nitrogen fixation activity by increased CO_2 concentration has been related to photosynthesis rather than a direct effect on the nitrogen fixing system.[49] As in other C_4-plants, photosynthesis in sugar cane should not be greatly affected by CO_2 concentrations.[25] The higher nitrogen fixation observed at low oxygen tension as compared with high tension was in accordance with previous findings[20], considering the fact that aerobic bacteria were removing O_2 to the advantage of semi-anaerobic N_2-fixers.

Work reported elsewhere in these Proceedings using both ^{15}N-labeled nutrient solution, [15]N_2 enriched gas, and acetylene reduction activity has demonstrated that surface sterilized and uninoculated "seed pieces" can, following "germination", develop the capacity to support N_2-fixation in culture solution.[43,48]

III. THE SITE OF FIXATION

Initial work[35] indicated that the site of nitrogen fixation seemed to be the roots, as fixed ^{15}N was present in separated roots of L64-43, but absent from separated shoots. Direct microscopic observations under the light microscope of roots stained with a cotton blue-lactophenol solution revealed that many of the epidermal cells were heavily colonized by microorganisms, some of them presumably involved in the fixation process. Later work[42] showed that in addition to [15]N_2 being fixed by organisms either in or on the root there was a very active dinitrogen fixation system present in the disturbed soil of the rhizosphere.

The incidence of aerobic and anaerobic N_2-fixing bacteria is high in all root systems of sugar cane grown in the field and is disturbed throughout different depths, down to 120 cm, and are present in the three main parts of cane root system: superficial, buttress, and rope roots.[37] Radiorespirometry studies reported elsewhere in this volume[15] show that the activity of the soil microbial population is strongly influenced by the presence of, and distance from, sugar-cane roots. This clearly points to the exudation of carbohydrate and/or growth factors from the root. Such soil microbiological activity is much higher for sugar-cane roots than for the roots of *Phaseolus* for example on the same soil type.

The presence of nitrogenase activity in germinated cuttings of sugar cane has already been noted in this paper.[36] This implies the presence of organisms capable of fixing dinitrogen. Sugar cane is reproduced by "germinating" pieces of node cut from mature stems stripped of leaves. These nodal stem pieces if tested for nitrogenase activity at the time of cutting give negative results, but if placed in culture solution and allowed to germinate, i.e., develop a shoot and adventitious roots at the node, then show strong nitrogenase activity. As far as is presently known, there is no nitrogenase activity in the stems of growing plants. Presumably the capacity for nitrogenase activity shown by germinating seed pieces and young seedlings is due to the removal of an inhibitor.

Subsequent ^{15}N experiment with seedlings have shown a high concentration of ^{15}N in the seed piece, initially higher than in roots or leaves.[43] Patriquin et al.[31] have demonstrated the presence of bacteria in the cells of seed pieces and particularly at the nodes. It can be supposed that the high concentration of carbohydrates in the "seed" stalk supports the energy required by N_2-fixing bacteria at the plant's seedling stage.

Work carried out by Arias et al.[5] compared electron microscope photographs taken from pure cultures of *Azotobacter* and *Azospirillum*, and bacteria washed from the root surface in order to compare with photographs of sections of normally grown roots incubated in N-free media and which showed high nitrogenase activity. Cells were observed in the case of the incubated root, filled with bacteria and spore-like bacteria, similar to bacteria outside the roots. The bacteria appeared to be primarily *Azotobacter* but there were others which were probably *Azospirillum.*

The presence of bacteria in the root, even of nitrogen-fixing types, does not prove that they are fixing nitrogen. Tritiated acetylene has been used in the determination of acetylene reduction activity (ARA) combined with electron microautoradiography[45] in an attempt to answer this question. The results suggested that fixation can occur within the root.

Patriquin et al.[31] found that there were high concentrations of bacteria around elongated ruptures at the base of roots newly developing from the seed piece. Subsequent work[54] showed that if surface sterilized seed pieces were germinated in sterile vermiculite, then the vermiculite subsequently showed ARA activity showing that bacteria had passed from the roots to the soil medium.

Fixation occurs in seedlings grown in solution culture, without inoculation and without solid medium.[43,48] Work therefore suggests that fixation may occur both in the root and external to the root, and in the case of young plants in the stem seed piece. The relative importance of root and rhizosphere soil-fixation cannot yet be evaluated, but Ruschel et al.[42] found that there was substantially more ¹⁵N fixed in rhizosphere soil than by the roots of 3-month old plants. The rhizosphere soil is likely to be of increasing importance with older plants and particularly with the rattoon crop about which we know little. The radiorespirometry studies represent the first work with the rattoon crop.

IV. CHARACTERISTICS OF THE N₂-FIXING ORGANISMS

By using N-free media the development of N-fixers can be easily observed by the incubation of 1 cm long root-sections, on surface media. Colonies are generally composed of different bacteria (*Azotobacter, Beijerinckia, Caulobacter, Vibrio*), and if the media is supplemented with calcium carbonate *Azotobacter* spp. are predominant. After 3 days incubation it is common to notice gas bubbles trapped under bacteria colonies, and there is evidence of saccharolitic *Clostridia* being present with high nitrogenase activity.[42]

Attempts to isolate N-fixing bacteria from the initial colony that flows from the cut surface of roots previously surface sterilized by 3 min under hypochlorite solution (10%) were initially not successful as evidenced after dilution by small colonies with slow growth, and no nitrogenase activity. Subcultures, of colonies different from *Azotobacter*, in sucrose or glucose N-free media did not show nitrogenase activity either. In semisolid malate-media, a small percent of isolates might develop a visible pellicle 2 to 4 mm below the surface, indicating *Azospirillum* like organisms, but which checked under 10% acetylene did not show nitrogenase activity. However, when 5 mℓ of yeast extract was added 80% of these isolates showed high activity of nitrogenase. More specifically, isolates on N-free media supplied with 2 ppb cobalt also showed growth and nitrogenase activity. These findings indicated that growth factors are needed for these bacteria. The requirement for cobalt is an interesting link with its requirement by other nitrogen fixing systems such as legume *Rhizobium*[1,10,19] and for *Azotobacter,*[28] and in *Alnus* and *Casuarina.*[7]

Aerobes and microaerophilic microorganisms influenced N₂-fixation of five different sugar-cane varieties, as indicated by nitrogenase activity.[36] The rate of ethylene evolution decreased in the fourth hour in all varieties, with the exception of CB 46-47 under normal and low oxygen atmosphere and NA 56-79 under low oxygen. In these two varieties, the microorganisms still appeared to be active after 4 hr of incubation under acetylene.

There is evidence of greater nitrogenase activity of *Azospirillum* in mixtures with bacillus-like non N-fixers, *Escherichia coli* and *Rhodotorula muscilaginosa* which showed greater nitrogenase activity than axenic cultures for the same level of inoculum and time incubation.[17] Whether this is due to non N_2-fixers supplying a growth factor or is due to their high respiration rate decreasing O_2 in the medium is not yet known.

V. THE EFFECT OF INOCULATION

It is clear that sugar cane has the capacity to perpetuate its bacterial system through the normal vegetative method of reproducing commercial cane, without further inoculation. It might therefore be suspected that inoculation would be superfluous. This is not necessarily so, because inoculation might to some extent overcome the delay in building up an appropriate bacterial population, while helping to ensure a high proportion of N_2-fixers in the rhizosphere. Moreover the numbers of bacteria in stalks are very variable[31] as is also the nitrogenase activity of seedlings, while bacterial numbers can also vary with season.[9] Inoculation should therefore result in more uniform and regular N_2-fixing capacity, less dependent on the initial stem population.

Benefits of inoculation of *Gramineae* by mixed cultures had previously been noted by Ruschel and Döbereiner[34] on rice, while Ruschel and Britto[53] demonstrated a high population of *Beijerinckia* in the rhizosphere of different *Gramineae* following inoculation. In Mahrashtra, India[24] inoculation of sugar cane with *Azotobacter* has shown some notable yield increases. Field trials over 3 years showed an increase of 50% (from 71 to 103 t/ha[1]) when the culture was smeared on the rootband of the seed piece at planting.

Also with sugar cane[38] replicated pot treatments were inoculated with a mixture of free-living N-fixers isolated from sugar-cane roots (*Azotobacter, Beijerinckia, Baccilus, Clostridium*). Results of an analysis of variance showed that an inoculation effect appeared 1 month after planting and disappeared 3.5 months later. The explanation of this is probably that the inoculum population initially had preferential access to root sites, but after 3.5 months the microorganism populations of both inoculated and noninoculated treatments became stabilized due to the subsequent increase in the free-living population.

Plants were however influenced by the apparent N_2-fixing activity of microorganisms only 3 months after planting, by which time roots were well distributed in the pot. The significant effect of inoculation observed at 1 and 2 months showed the main effect when cane leaves were added to the soil, indicating that sources of energy stimulate the development of N_2-fixers and that organic nitrogen added by leaves did not affect N_2-fixers activity. Evidence for this was supported by results from noninoculated treatments 3 months after planting. In this case the cane planted in soil supplemented with cane leaves did not show any response, indicating the favorable effect, at least in the short term, of bacteria selected for survival and efficiency in these conditions.

Recent work in solution culture[43] has demonstrated increased ARA and more uniform response following inoculation. It is a common commercial practice when establishing ''nurseries'' of sugar cane to treat the seed pieces in hot water (50°C) as a disease prevention measure. When seed pieces were treated in such a manner and grown in solution culture, ARA was very small. This suggests that inoculation following heat treatment could be a useful agronomic practice.

VI. INFLUENCE OF VARIETY

Nitrogenase activity was estimated in roots, germinated cuttings, whole plants of sugar cane (2-months old) in intact and disturbed root systems, under low and normal (air) oxygen atmosphere.[36] The results showed evidence of a varietal effect on N_2-fixing (ethylene) activity of microorganisms in sugar cane. Varieties NA56-62 and CB46-47 had on average the greatest nitrogenase activity, which in the intact system (normal oxygen atmosphere) differed statistically from the others. With low oxygen atmosphere CB47-355 also showed high nitrogenase activity, while variety CB41-76 had the lowest nitrogenase activity in all experiments. It may be noted that varieties CB46-47 and NA56-62 are high in sucrose content of mature stalks, while CB41-76 has a low sucrose content, although having good productivity.[6]

Although earlier work had indicated[35] that the sites of N-fixation were in the roots, this later work showed that germinated cuttings of some varieties from which the roots and shoots were excised, had a higher nitrogenase activity than the rest of the plant parts. There appear to be differences (not statistically significant) in nitrogenase activity among the internal part of the cuttings, the external part (rind) and the roots, with the rind showing the highest activity.

Further work reported here[44] has looked at the various ancestral lines from which the above varieties were derived. The results point towards the possibility of selecting for nitrogen fixation potential. Plant breeding is probably the only way to increase nitrogen fixation in sugar cane, as the bacterial system appears to be too complicated to achieve improved efficiency through its modification.

VII. NITROGEN FIXATION UNDER FIELD CONDITIONS

Sugar cane root samples taken from the field will usually show ARA. Determining on a field scale the capacity of the associative N_2-fixation system of sugar cane to fix atmospheric nitrogen is clearly very important, but there are many practical difficulties to be overcome. From the experimental point of view, major difficulties are the length of the growing season (over 12 months), the large size of the plants, the potential great dilution of ^{15}N tracer experiments, the difficulty of finding an appropriate nonfixing test plant for ^{15}N tracer experiments, the wide field spacing which requires a large amount of ^{15}N fertilizer, the inhibiting effect of NH_4 fertilizer,[39,47] and the rapidity with which labeled fertilizer moves down the profile under Brazilian conditions.[8]

A study examined the effect of N-fertilization and irrigation on nitrogenase activity in the field.[39] The nitrogenase activity of the sugar-cane roots was low during the initial development, but after 12 and 15 months up to a 100-fold increase was observed. This behavior may probably be attributed to the influence of temperature on the development of the population of N_2-fixing microorganisms, and the nature of root exudates, such as the enhancement of carbohydrates during plant development.

The nitrogenase activity of roots samples after 6 and 9 months was higher in irrigated than nonirrigated plants at all fertilizer levels. According to Evans, cited by van Dillewijn,[46] the amount of root exudate decreases during drought but after copious watering root exudation resumes vigorously. This may explain the differences observed for irrigated plants. Twelve months after planting the highest nitrogenase activity was found in nonirrigated plant roots. After 15 months the difference between irrigated and nonirrigated plants was insignificant and it is suggested that this behavior might be an effect of the abundant rainfall in January and April. Kishan Singh[55] suggests that the beneficial effect of *Azotobacter* inoculation found in Mahrashtra is directly related to the high frequency of irrigation, about 30 times per year.[39]

The nitrogenase activity of irrigated plants decreased with increasing ammonium sulphate fertilization. In nonirrigated plots nitrogen fertilization did not seem to affect nitrogenase activity. Variations of phosphate and potassium fertilization did not appear to affect nitrogenase activity at any level investigated.

A theoretical extrapolation of potential N_2-fixation was made by Ruschel et al.[42] as follows: assuming a yield of 100 t/ha, and a top/root ratio of 15:1 in planted sugar cane[23] and a relatively linear growth rate of the root system, the roots of sugar cane could fix 0.2 to 0.38 and 0.44 to 23.8 g of N/ha/day based on the values of ^{15}N uptake by nonamended, intact root system placed under simulated normal atmospheric conditions, and disturbed systems, respectively. A similar calculation of fixed nitrogen taken up by sugar cane leaves in an intact system yields values of 5.7 to 16.9 g of N/ha/day. On the other hand ^{15}N fixed in the soil, as tested in a disturbed system, amounted to 250 g N/ha/day, assuming a soil density of 1.3, a 20 cm soil layer where roots are most effective, and a linear rate of root growth. Assuming a fixation period of at least 200 days in the life of the crop this would amount to 3.38 kg for "plant" fixation and 50 kg for rhizosphere N_2-fixation. This corresponds to about 30% of the nitrogen required by an average crop.

In fact the effect of crop growth stage on N_2-fixation is not known, nor is the proportion of days during the growth period on which significant fixation takes place. It is known to be reduced by drought. It could well be reduced during periods of extremely rapid growth, when it is likely that all the plant's carbohydrate production is required to sustain the new development. Thus Matsui et al.[27] carried out a field experiment with [15]N_2 enriched atmosphere over a 5 day period, which coincided with the very rapid elongation stage of 9 month old plant cane, and detected ^{15}N enrichment of the soil taken from close to both big and small roots. However, ^{15}N enrichment was not detected in leaves or stems, possibly due to dilution effect.

In default of developing satisfactory ^{15}N-fertilizer experiments we have tried using natural isotopic variation ($\delta°/oo$ [15]N) to obtain an integrated value for fixation. The high $\delta°/oo$ values usually found in Brazilian soils make this possible, but the method has the inherent limitations associated with soil and plant sampling, and the fact that it takes only a small difference in ^{15}N/^{14}N ratio to indicate a comparatively large amount of fixation. Applying this technique to a N-balance experiment[40] carried out in large containers (90 kg soil) indicated that about 17% of the plant nitrogen at harvest was due to fixation.[41] With a field test on 8-month old plant cane it was found (Table 1) that six varieties varied between 5.76 and 7.35 $\delta°/oo$ [15]N, the differences being significant, compared with a soil δ^{15}N value of 10.73%.[47] The percent of plant-N derived from fixation calculated on these figures comes to over 30%. Such amounts may appear high, but if N_2-fixation is responsible in Brazil for the relative lack of response of sugar cane to N-fertilization, then such high figures would be necessary.

VIII. DISCUSSION

Better information on the level of field N_2-fixation is needed, because at some future time important decisions may have to be made in relation to plant breeding. Most current sugar-cane varieties in Brazil are "old", moreover it is not known what effect the support of a N_2-fixing association has on sugar production and yield. It could well be that newer high yielding varieties selected for better N-response and grown with high N-fertilization levels will not have the ability to support associative N_2-fixation. Under some conditions of culture it might be more profitable to have varieties with poor capacity for N_2-fixation but with good response to N-fertilizer. However, with the energy problem now acute and the high cost of fertilizer it might be better to sacrifice ultimate yield for good nitrogen fixing capacity, but at present we do not have good figures to give to economists and plant breeders. With about two million hectares of sugar cane in Brazil,[4] 10% of the world crop, the amounts of nitrogen involved are potentially great.

Table 1
δ^{15}N VALUES FOR 8-MONTH OLD PLANTS OF FIVE SUGAR-CANE VARIETIES

Variety	δ^{15}N value
NA56-62	6.38
CB41-76	5.76
CB41-355	6.19
IAC51/205	6.95
CP51-22	7.35
Soil δ^{15}N = 10.73	
Sig. diff. ($P = 0.05$)	0.27

From Vose, P. B., Ruschel, A. P., and Salati, E., *2nd Latin American Botanical Congr., Brasilia,* Abstr. 89, 90, January 22 to 29, Brazil, 1978. With permission.

So far there is no evidence as to whether the N$_2$-fixing system is common to all sugar-cane varieties, or to all areas. It appears to have a number of different bacterial types, not all of which fix nitrogen. Possibly there may be a certain adventitious element in the mixed colonies depending on the nature and relative numbers of free-living soil organisms in any situation, or on the plant genotype. Probably a loose association of a number of forms is involved. The close symbiotic nature of the association is indicated by the fact that if system bacteria hitherto actively fixing are subcultured, then they may cease N$_2$-fixation as pure cultures, to regain the capacity to fix when supplied with growth factors. Whether in normal circumstances the mixed colony receives these factors from the sugar-cane roots or from a dominant member of the colony is not yet known.

The extent of an associative N$_2$-fixing system in sugar cane outside Brazil is not known. It could be that selection for high nitrogen response and the application of high levels of NH$_4$-fertilizer has eliminated it with highly developed agronomy. There are many reports of certain varieties showing lack of response to N-fertilizer in nearly all sugar cane growing areas, and equally, varieties which do respond. It is apparent that work is now necessary on examining the correlation between variety response to N-fertilization and the capacity for associative N-fixation. Additionally we need to have greater knowledge of the basic yield physiology of sugar cane in relation to its utilization of nitrogen in the production of dry matter.

It seems apparent that if it is desirable to increase the efficiency of associative N-fixation then it must almost certainly be done through plant breeding rather than through modifying the bacterial system, which is probably too complicated to make this approach viable. Approaching increased effectiveness of associative N-fixation through sugar-cane breeding seems a definite possibility on the results so far obtained.

There is a need to clarify the plant physiological factors associated with N$_2$-fixation in sugar cane. For example, the effect of light, temperature (the effectiveness of the system is known to be greatly reduced at low temperatures), day length, sugar formation, etc. We need to know the biochemistry which determines the capacity or lack of capacity of a sugar-cane cultivar for N$_2$-fixation.

From the standpoint of practical agronomy of sugar cane there is a need to determine if modification of cultural practices, cultivation, fertilization, inoculation, irrigation, etc. can be used to increase associative N$_2$-fixation of existing commercial varieties on a field scale.

REFERENCES

1. **Ahmed, S. and Evans, H. J.**, Cobalt a micronutrient element for the growth of soybean plants under symbiotic conditions, *Soil Sci.*, 90, 205, 1960.
2. **Alvarez, R., Segalla, A. L., and Catani, R.**, Adubação nitrogenada de cana-de-açucãr. III. Fertilizantes nitrogenados, *Bragantia*, 16, 23, 1957.
3. **Anon.**, Planalsucar Annual Report for 1975, Piracicaba, São Paulo. Brasil, 1976.
4. Anũario Estaẗistico do Brasil, Secretaria de Planejamento da Presidência da República, Fundação Inst. Brasil, *Geogr. Esẗatĩstica*, 1976.
5. **Arias, O. E., Gatti, Irene M., Silva, D. M., Ruschel, Alaides P., and Vose, P. B.**, Primeras observaciones al microscõpio eletrônico de bacterias fijadoras de N_2, en la raiz de caña de azúcar (*Saccharum officinarum*, L.). *Rev. Turrialba*, 28, 203, 1978.
6. **Bassinello, A. I.**, Apreciação sobre experimentos de competição de variedades da Série 71, *Bras. Açucareiro*, 87(5), 42, 1976.
7. **Bond, G. and Hewitt, E. J.**, Cobalt and the fixation of nitrogen in root nodules of *Alnus* and *Casuarina, Nature (London)*, 195, 94, 1962.
8. **Cervellini, A., et al.** Dynamics of Nitrogen in an Intensive Plot Planted to Beans, Rep. Res. Contract 1597, Working paper No. 34 presented to 4th Research Coordination Meeting FAO/IAEA/GSF Coordinated Programme on Agricultures Nitrogen Residues, Piracicaba, Brasil, 1978.
9. **Costa, J. M. F. and Ruschel, A. P.**, Seasonal variation in the microbial population of sugar cane stalks, in *Associative N_2-Fixation*, Vose, Peter B. and Ruschel, A. P., Eds., CRC Press, Boca Raton, Fla., 1981.
10. **Delwiche, C. C., Johnson, C. M., and Reisenauer, H. M.**, Influence of cobalt on nitrogen fixation by *Medicago, Plant Physiol.*, 36, 73, 1961.
11. **Döbereiner, J.**, Influência da cana-de-açúcar na população de *Beijerinckia* do solo, *Rev. Bras. Biol.*, 19, 251, 1959a.
12. **Döbereiner, J. and Alvahydo, R.**, Sobre a influencia da cana-de-açúcar na ocorrência de *Beijerinckia* no solo. II. Influência das diversas partes do vegetal, *Rev. Bras. Biol.*, 19, 401, 1959b.
13. **Döbereiner, J., Day, J., and Dart, P. J.**, Nitrogenase activity in the rhizosphere of sugar cane and some other tropical grasses, *Plant Soil*, 37, 191, 1972a.
14. **Döbereiner, J., Day, J., and Dart, P. J.**, Rhizosphere associations between grasses and nitrogen-fixing bacteria: effect of O_2 and nitrogenase activity in the rhizosphere of *Paspalum notatum, Soil Biol. Biochem.*, 5, 157, 1972b.
15. **Freitas, J. R., Ruschel, A. P., and Vose, P. B.**, Radiorespirometry studies as an indication of soil microbiological activity in relation to the root system in sugar cane, and comparison with other species, in *Associative N_2-Fixation*, Vose, Peter B. and Ruschel, A. P., Eds., CRC Press, Boca Raton, Fla., 1981.
16. **Gomes, F. P. and Cardoso, E. M.**, *Adubação Mineral da Cana-de-Açúcar*, Editora Aloisi Ltda, Piracicaba, Brasil, 1958, chap. 7.
17. **Graciolli, L. A. and Ruschel, A. P.**, Atividade da nitrogenase de *Spirillum* sp. em cultura pura e em misturas com microrganismos não fixadores de nitrogênio, *Rev. Bras. Ci. Solo*, 2, 215, 1978.
18. **Graciolli, L. A. and Ruschel, A. P.**, Microorganisms in the phyllosphere and rhizosphere of sugar cane, in *Associative N_2-Fixation*, Vose, Peter B. and Ruschel, A. P., Eds., CRC Press, Boca Raton, Fla., 1981.
19. **Hallsworth, E. G., Wilson, S. B., and Greenwood, E. A. N.**, Copper and cobalt in nitrogen fixation, *Nature (London)*, 187, 79, 1960.
20. **Hardy, R. W. F. and Havelka, U. D.**, Nitrogen fixation research: a key to world food, *Science*, 188, 633, 1975.
21. **Harris, D. and Dart, P. J.**, Nitrogenase activity in the rhizosphere of *Stachys sylvatica* and some others dicotyledoneus plants, *Soil Biol. Biochem.*, 5, 277, 1973.
22. **Hartt, E. E., Kortschak, H. P., Forbes, A. J., and Burr, G. O.**, Translocation of ^{14}C in sugar cane, *Plant Physiol.*, 39, 305, 1963.
23. **Humber, R. P.**, *The Growing of Sugar cane*, Am. Elsevier, New York, 1963, chap. 1.
24. **Jadhav, J. S. and Andhale, S. S.**, Biological nitrogen fixation in sugar cane with specific reference to *Azotobacter, Sugar News*, 8(4), 8, 1976.
25. **Kortschak, H. P., Hartt, C. E., and Burr, G. O.**, Carbon dioxide fixation in sugar cane leaves, *Plant Physiol.*, 40, 209, 1965.
26. **Malavolta, E., Haag, H. P., Mello, F. A. F., and Brasil Sobo., M. O. C.**, *Nutrição Mineral de Plantas Cultivadas*, 1974, chap. 5.
27. **Matsui, E., Vose, P. B., Rodrigues, N. S., and Ruschel, A. P.**, Use of ^{15}N enriched gas to determine N_2-fixation by undisturbed plants in the field, in *Associative N_2-Fixation*, Vose, Peter B. and Ruschel, A. P., Eds., CRC Press, Boca Raton, Fla., 1981.

28. Nicholas, D. J. D., Kobayasti, M., and Wilson, P. K., Cobalt requirement for nitrogen metabolism in microorganisms, *Proc. Nat. Acad. Sci. U.S.A.*, 49, 1962.

29. Orlando Fo., J. and Haag, H. P., InfluÇencia Varietal e do Solo no Estado Nutricional na Cana-de-Açúcar (*Saccharum* spp.) Pela Análise Foliar, Boletim Técnico No. 2, Coordenadoria Regional Sul PLANALSUCAR, Araras, 1976, 52.

30. Patriquin, D. and Knowles, R., Nitrogen fixation in the rhizosphere of marine angiosperms, *Mar. Biol.*, 16, 49, 1971.

31. Patriquin, D., Graciolli, L. A., and Ruschel, A. P., Nitrogenase Activity and Bacterial Infection of Germinated Sugar Cane Stem Cuttings, Steenbock-Kettering Int. Symp. on Nitrogen Fixation, Madison, Wis., 1978.

32. Ruschel, A. P., Fixação Biológica de Nitrogênio em Cána-de-Açúcar (*Saccharum* spp.), Anais 2nd Congr. Latino-Americano de Botânica e XXVI Congr. Nacional de Botânica, Brasília, Brazil, 1978.

33. Ruschel, A. P., Fixacão Biológica de Nitrogênio, in *Fisilogia Vegetal*, M. G. Ferri, Ed., 1979, chap. 4.

34. Ruschel, A. P. and Döbereiner, J., Bactérias Assimbióticas Fixadoras de Nitrogênio na Rizosfera de Gramíneas Forrageiras, Boletim No. 7 IPEACS-DPEA, Min. Agric., 1966.

35. Ruschel, A. P., Henis, Y., and Salati, E., Nitrogen-15 tracing of N-fixation with soil-grown sugar cane seedlings, *Soil Biol. Biochem.*, 7, 181, 1975.

36. Ruschel, A. P. and Ruschel, R., Varietal differences affecting nitrogenase activity in rhizosphere of sugar cane, *Proc. Int. Soc. Sugar Cane Technol.*, 2, 1941, 1977a.

37. Ruschel, A. P., Orlando Fo., J., Zambello Jo., E., and Henis, Y., Aerobic and anaerobic nitrogen-fixing bacteria on sugar cane roots, *Proc. Int. Soc. Sugar Cane Technol.*, 2, 1923, 1977b.

38. Ruschel, A. P., Orlando Fo., J., Zambello Jo., E., and Henis, Y., Aerobic and anaerobic N-fixing microorganisms in soil-grown sugar cane as affected by cane plant, N-fixers inoculation and soil addition of cane leaves, *Proc. Int. Soc. Sugar Cane Technol.*, 2, 1923, 1977c.

39. Ruschel, A. P., Orlando Fo., J., and Zambello Jo., E., The effect of nitrogen, phosphorus and potassium fertilization and irrigation on nitrogenase activity and yield of sugar cane, *Proc. Int. Soc. Sugar Cane Technol.*, 2, 1903, 1977d.

40. Ruschel, A. P., Matsui, E., Orlando Fo., J., and Bittencourt, V. C., Closed system balance studies in sugar cane utilizing ^{15}N-ammonium sulphate, *Proc. Int. Soc. Sugar Cane Technol.*, 2, 1539, 1977e.

41. Ruschel, A. P. and Vose, P. B., Present Situation Concerning Studies on Associative N$_2$-Fixation in Sugar Cane, *Saccharum officinarum*, Boletim Científico BC-045, Centre de Energia Nuclear na Agricultura, Piracicaba, Brasil, 1977.

42. Ruschel, A. P., Victoria, R. L., Salati, E., and Henis, Y., Nitrogen fixation in sugarcane (*Saccharum officinarum*), *Ecol. Bull. Stockholm*, 26, 297, 1978.

43. Ruschel, A. P., Matsui, E., and Vose, P. B., and Salati, E., Potential N$_2$-fixation by sugar cane *Saccharum* sp. in solution culture. II. Effect of inoculation; and dinitrogen fixation as directly measured by ^{15}N$_2$, in *Associative N$_2$-Fixation*, Vose, Peter B. and Ruschel, A. P., Eds., CRC Press, Boca Raton, Fla., 1981.

44. Ruschel, R. and Ruschel, A. P., Inheritance of N$_2$-fixing ability in sugar cane, in *Associative N$_2$-Fixation*, Vose, Peter B. and Ruschel, A. P., Eds., CRC Press, Boca Raton, Fla., 1981.

45. Silva, D. M., Ruschel, A. P., Matsui, E., Nogueira, N. L., and Vose, P. B., Determination of the activity of N$_2$-fixing bacteria in sugar cane roots and bean nodules using tritiated acetylene reduction and electron microautoradiography, in *Associative N$_2$-Fixation*, Vose, Peter B. and Ruschel, A. P., Eds., CRC Press, Boca Raton, Fla., 1981.

46. van Diilewijn, C., *Botany of Sugarcane*, Chronica Botanica, Waltham, Mass., 1952, 371.

47. Vose, P. B., Ruschel, A. P., and Salati, E., Determination of N$_2$-Fixation, Especially in Relation to the Employment of Nitrogen-15 and of Natural Isotope Variation, 2nd Latin American Botanical Congr., Brasilia, Abstr. 89 to 90, January 22 to 29, Brazil, 1978, in press.

48. Vose, P. B., Ruschel, A. P., Victoria, R. L., and Matsui, E., Potential N$_2$-fixation by sugar cane *Saccharum* sp. in solution culture. I. Effect of NH + 4, NO$_3$, variety and nitrogen level, in *Associative N$_2$-Fixation*, Vose, Peter B. and Ruschel, A. P., Eds., CRC Press, Boca Raton, Fla., 1981.

49. Wilson, P. B., *The Biochemistry of Symbiotic Nitrogen Fixation*, University of Wisconsin Press, Madison, 1940, 114.

50. Yoshida, T. and Ancajas, R. R., Nitrogen fixation by bacteria in the root zone of rice, *Soil Sci. Soc. Am. Proc.*, 35, 156, 1971.

51. Takahashi, D. T., ^{15}N field studies with sugar cane in coral soils, *Hawaii. Plant. Rec.*, 58, 119, 1970.

52. Arruda, H. V., Adubacão nitrogenada na cana-de-açúcar, *Bragantia*, 19, 1105, 1960.

53. Ruschel, A. P. and Britto, D. P. P. S., Fixaçao assimbiótica de nitrogênio atmosférico em algumas gramíneas e na tiririca pelas bacterias do gênero *Beijerinckia* Derx, Bol. Téc. 13, IPEACS, Rio de Janeiro, Brasil, 1966.

54. Ruschel, A. P., unpublished.

55. Singh, Kishan, personal communication.

Chapter 13

MICROORGANISMS IN THE PHYLLOSPHERE AND RHIZOSPHERE OF SUGAR CANE*

L. A. Graciolli** and A. P. Ruschel***

TABLE OF CONTENTS

*   Translator: Diva Athié.
**   Faculty of Agronomy of Paraguacu Paulista (ESAPP), São Paulo, Brazil.
***Centro de Energia Nuclear na Agricultura, Piracicaba, São Paulo, Brazil.

I. INTRODUCTION

Much work has been carried out since Döbereiner[6,7] first observed the association root/N$_2$-fixing bacteria in sugar cane, and studies on this association have increased considerably in the last few years. Recent results have indicated the existence of bacteria in the interior of stalks used for sugar cane vegetative propagation, which are concentrated in the periphery of the stem nodes, and after germination populations of bacteria appear in the rhizoplane and the rhizosphere.[12]

Research on biological nitrogen fixation in the phyllosphere is also recent. First observations[15] showed common occurrence of *Beijerinckia* sp. on the surface of leaves, in Java and Sumatra (Indonesia). Other investigations have been described, but with *Gramineae* and specifically sugar cane, very little has been done.

The objective of the present work was to study the presence of bacteria in the leaves and the interior of stems and associated with the rhizosphere.

II. MATERIAL AND METHODS

Leaves both from seedlings grown in pots and collected in the field were used to study sugar cane phyllosphere and to study the rhizosphere of roots germinated from seed stalks or setts. These setts were comprised of a piece of mature stem containing a node with nodal bud, and were about 5 cm long.

A. Phyllosphere Studies

Leaves of 30-day old sugar cane seedlings from Planalsucar, Araras, S.P., of three varieties CO 513, IAC 48-65, and CB 50-41, were separated (0.5 g) and received two treatments: with and without sterilization. Surface sterilization was made with sodium hypochlorite (Q-Boa at 20%) for 20 min and washed four times in sterile water. Dilutions were then made for each treatment, which were used as inoculum in three different culture media (9 mℓ), distributed in 20 mℓ flasks.

1. LG medium: K$_2$HPO$_4$ — 0.1 g; KH$_2$PO$_4$ — 0.4 g; FeCl$_3$ — 0.01 g; MgSO$_4$·7H$_2$O — 0.02 g; NaMoO$_4$·2H$_2$O — 0.002 g; NaCl — 0.1 g; bromothymol blue alcohol solution at 0.5% - 5 mℓ; saccharose — 10.0 g; agar — 12 g; distilled water —
2. LGY medium: same as LG plus 0.5 g yeast extract/ℓ.
3. LG-Malate medium.[4]

After inoculation, the tubes were incubated for 4 to 5 days at 30°C and nitrogenase activity was determined by the acetylene reduction method,[3] with a Beckman® GC65 gas chromatograph using an H$_2$ flame ionization detector at 175°C and a glass column with internal diameter 1/8 in. × 1.60 m containing Porapak N, 8-100 mesh, at 110°C.

For the field studies, young leaves were collected from 11-month old field plants of the same varieties, being cut close to the stem. Treatments were as previously described.

B. Rhizosphere Studies

Sugar cane-stalks from the field (Planalsucar, Araras, S.P.) and from approximately 12-month old plants grown in the greenhouse of varieties NA and CB, were cut into setts and numbered from the root, and surface sterilized in alcohol (95%) for 10 min and sodium hypochlorite (Q-Boa 20%) for 20 min, with successive washings in sterile water and sterile phosphate buffer (0.05 *M*, pH 7.0). They were germinated in 500 mℓ flasks containing 250 mℓ sterilized and moistened vermiculite, and covered with cotton. The flasks were then taken to the greenhouse and whenever necessary the plants were watered with sterile water for the first 10 days and from then on with a sterile nutrient solution (N-free).

Nitrogenase activity was determined and decimal dilutions of stem, root and rhizosphere soil were made when no nitrogenase activity was observed. Culture medium used was LGP.[12]

1. Plants in Nutrient Solution

Sugar-cane stalks of varieties NA 56-79 and CB 41-76 were germinated in the greenhouse in 20 l plastic pots after surface sterilization; 0.35 and 14 ppm $^{15}N_2$ as NO_3^- were added to the nutrient solution.[20] Sixty days after germination, 1 g root was taken from the different treatments for decimal dilutions, as was done with the phyllosphere material, in three culture media: (1.75 g/l agar), LG, LGY, LGP.

2. Root Pieces

Pieces of root were collected from the different treatments both close to the stem and from the ends and distributed in Petri dishes containing, separately, the culture media LG and LGY. Nitrogenase activity was determined as before. The pieces of root were placed individually together with the culture medium in 10 ml flasks.

3. Incubation of Roots in Tetrazolium

Pieces of roots both from the plants grown in vermiculite and in nutrient solution were incubated in tetrazolium solution (sterile phosphate buffer, 0.05 M, pH 7.0 and 1% glucose plus 1.5 g/2,3,5 triphenyltetrazolium chloride (TTC) for one night.[12] After incubation root pieces or transverse sections were prepared with the Hooker® (Lab Line Instrument Co.) microtome and observed by light microscope.

III. RESULTS AND DISCUSSION

Table 1 shows results from the phyllosphere experiments and Tables 2, 3, and 4 and Figures 1, 2, and 3 show results of the rhizosphere experiments.

A. Phyllosphere

There is no doubt that there is a large number of N_2-fixing microorganisms in the leaves of sugar cane (Table 1). It is possible that some species were in the vascular tissue of the leaves, identification tests[18] on these indicated *Azospirillum brasilense* (variety CO 513, whose young leaves had been sterilized). As these bacteria do not form cysts and were isolated from a surface sterilized leaf, there is little possibility that they were on the surface of the leaf. Nitrogen fixing bacteria were found in the interior of the stems and roots[12] (not identified). It is possible that transpiration contributed to the movement of bacteria in the leaves through the vascular tissue. It can be noted in Table 1, especially in seedling leaves, that the greatest positive dilutions for acetylene reduction activity (ARA) are in the LGY medium, indicating that growth factors are necessary for the normal metabolism of the bacteria. High positive dilution of dry leaves can also be noted, which means that the dry leaves can contribute to the maintenance of N_2-fixing microflora of this system when they fall to the ground.

Ruschel[16] observed a marked increase in the nitrogen content of the plant when cut cane leaves were added to the soil. It is true that this increases the C/N ratio and therefore there is a prompt increase in N_2-fixation. It seems now that N_2-fixing and nonfixing microorganisms were inoculated at the same time, and in the general context, the nonfixers can help to stabilize an optimum microhabitat for fixation.

Table 1
NITROGENASE ACTIVITY (N MOL C_2H_4/HR/FLASK) OF DECIMAL DILUTIONS OF LEAVES OF SEEDLINGS, YOUNG, AND DEAD LEAVES, IN DIFFERENT CULTURE MEDIA, OF THREE SUGAR-CANE VARIETIES, AFTER 4 DAYS INCUBATION AT 30°C

Variety	Treatment	Dilution	Leaves of seedlings			Young leaves			Dead leaves		
			LG	LGY	LGM	LG	LGY	LGM	LG	LGY	LGM
Co513	Sterilized surface	10^1	0	148.2	0	90.7	336.6	43.2	86.4	268.8	89.7
		10^2	0	95.8	0	16.8	84.9	0	139.2	92.6	0
		10^3	0	9.8	0	0	8.4	0	0	0	0
		10^4	—	—	—	0	0	0	0	0	0
		10^5	—	—	—	0	0	0	0	0	—
		10^6	—	—	—	—	—	—	—	—	—
	Nonsterilized surface	10^1	4.9	256.8	70.5	345.6	441.6	61.4	1075.0	468.4	
		10^2	4.1	331.9	152.1	113.9	104.2	0	622.0	128.6	122.8
		10^3	0	5.6	0	30.5	96.0	0	214.0	142.0	51.3
		10^4	—	—	—	0	0	0	28.8	263.0	60.4
		10^5	—	—	—	0	0	0	12.8	0	164.1
		10^6	—	—	—	—	—	—	0	—	—
IAC48/65	Sterilized surface	10^1	0	169.9	0	19.8	0	0	0	0	0
		10^2	0	65.7	0	0	0	0	136.8	313.9	0
		10^3	0	262.8	0	0	0	0	0	0	0
		10^4	—	—	—	0	0	0	0	0	—
		10^5	—	—	—	0	0	0	0	0	—
		10^6	—	—	—	—	—	—	—	—	—
	Nonsterilized surface	10^1	98.8	133.3	182.7	61.9	10.8	80.6	72.0	20.5	12.9
		10^2	0	163.0	3.7	14.4	3.6	25.9	187.2	216.0	0
		10^3	0	108.6	0	18.0	0	0	27.0	201.6	0
		10^4	—	—	—	0	0	0	20.0	239.0	0
		10^5	—	—	—	0	0	0	0	127.0	0
		10^6	—	—	—	—	—	—	0	—	—
CB50/41	Sterilized surface	10^1	0	0	0	59.2	253.4	0	25.9	40.8	76.8
		10^2	0	29.8	0	3.6	126.7	0	2.4	0	0
		10^3	0	138.3	0	0	74.4	0	7.0	0	0
		10^4	—	—	—	—	0	0	0	0	0
		10^5	—	—	—	—	0	0	0	—	—
		10^6	—	—	—	—	—	—	—	—	—
	Nonsterilized surface	10^1	0	339.8	0	103.0	556.8	0	135.3	576.0	34.0
		10^2	0	4.4	0	106.0	0	0	133.1	288.0	20.1
		10^3	0	4.6	0	0	0	0	154.5	12.7	44.4
		10^4	—	—	—	0	58.0	0	10.8	0	66.7
		10^5	—	—	—	0	0	0	0	0	0
		10^6	—	—	—	—	—	—	—	—	0

Note: Dash (—) = not analyzed.

The most common bacteria isolated was *Beijerinckia*. Other spore forming N_2-fixing bacteria were also isolated from the surface of the leaves. This can be explained by (1) spore resistance to hypochlorite, (2) there are in the phyllosphere microhabitats which protect the bacteria, similar to those found in the roots,[5] or (3) there are bacteria in the interior of the leaves. Fungi, yeast and other bacteria were also found.

B. Rhizosphere
Figure 1 shows that the ARA of plants obtained from cane stalks (varieties NA and CN) grown in the greenhouse, is concentrated in the stalks close to the roots (setts 3

Table 2

NITROGENASE ACTIVITY (ARA) OF DECIMAL
DILUTIONS, IN LPG MEDIUM, OF STALKS FROM
FIELD THAT DID NOT REDUCE C_2H_2 IN THE INTACT
SYSTEM WITHOUT AERIAL PARTS; AFTER 3 DAYS
INCUBATION AT 30°C

Stalk	Variety	Part assayed	Highest positive dilution that showed ARA	n mol of ethylene/ flask/hr/highest positive dilution
		Stalk	10^3	374
8	NA	Root	0	0
		Rhizosphere	10^3	78
		Stalk	10^3	409
15	NA	Root	10^1	22
		Rhizosphere	10^1	398
		Stalk	10^1	346
10	CB	Root	10^1	100
		Rhizosphere	10^1	292
		Stalk	0	0
18	CB	Root	0	0
		Rhizosphere	0	0

Table 3

ARA OF DECIMAL DILUTION OF SUGARCANE
ROOTS, CULTIVATED IN NUTRIENT SOLUTION,
WITH DIFFERENT LEVELS OF N (0,3,5 AND 14 PPM)
IN DIFFERENT CULTURE MEDIA

Variety	Treatment	Maximum dilution that showed ARA			Maximum dilution with any bacterial growth		
		LG	LGY	LGP	LG	LGY	LGP
NA	0	10^3	10^2	10^2	10^5	10^8	10^{10}
NA	3.5	10^5	10^3	10^2	10^6	10^{10}	10^{10}
NA	14	10^3	10^2	10^2	10^5	10^9	10^{10}
CB	0	10^2	10^1	10^2	10^5	10^7	10^{10}
CB	3.5	10^3	10^4	10^2	10^5	10^9	10^{10}
CB	14	10^3	10^1	10^2	10^5	10^7	10^{10}

and 4, respectively) and the most distant (sett 16). Figure 2 shows uniform distribution of ARA in setts of plants from the field. It is true that the weight of the greenhouse stalks was about half (10 g) that of those from the field, which might have influenced the number of bacteria moving to the rhizosphere after germination. It can be initially concluded that nitrogen fixing microorganisms were present only in ARA positive setts. However, Figure 3 shows ARA in different parts of the plant and the rhizosphere soil which showed no activity in the intact system (Figure 1), but did when stalks and roots were separated, suggesting the presence of N_2-fixing microorganisms in all parts analyzed, but which were inhibited for some unknown reason. It was noted that the foliar system of plants from the greenhouse was not as green and developed as those from the field, this probably influencing root exudation and consequently biological nitrogen fixation. Many authors[1,2,8,14] agree with this hypothesis, however, it should be considered that substances inhibiting N_2-fixation, if present in the interior of the stalks, could also be exuded to the rhizosphere and inhibit the bacteria. This possibility

Table 4
NITROGENASE ACTIVITY (μ MOLES
C₂H₄/HR/ROOT PIECE) OF ROOT
PIECES INCUBATED FOR 3 DAYS IN LG
AND LGY MEDIUM AT 30°C, THE
PLANTS HAVING BEEN CULTIVATED
IN NUTRIENT SOLUTION WITH OR
WITHOUT ADDITION OF N. EACH
VALUE IS THE MEAN OF THREE
REPLICATES

Culture medium	Zero N in the solution		+ N in the solution	
	Tip	Root close to stalk	Tip	Root close to stalk
LG	0.586	1.350	1.112	1.114
LGY	3.175	6.355	8.512	7.571

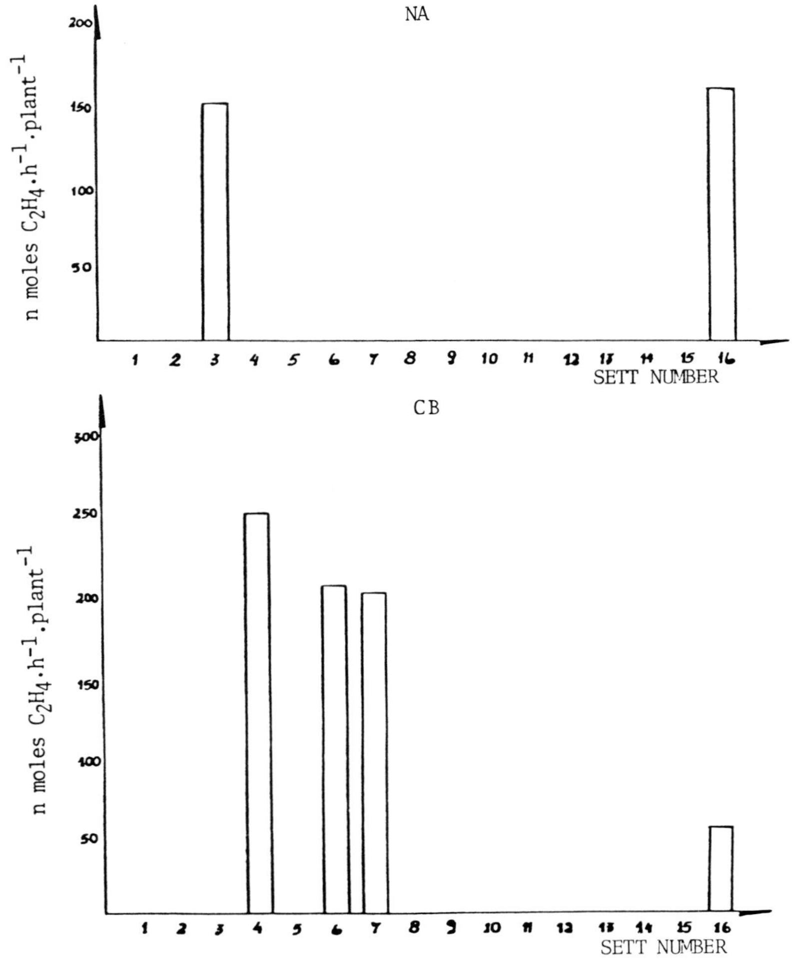

FIGURE 1.   Nitrogenase activity (n mol C₂H₄/hr) of setts of varieties Na and BC taken from the greenhouse, surface sterilized and aseptically planted in sterile vermiculite. Numbered from the root upwards. After 3½ weeks from planting.

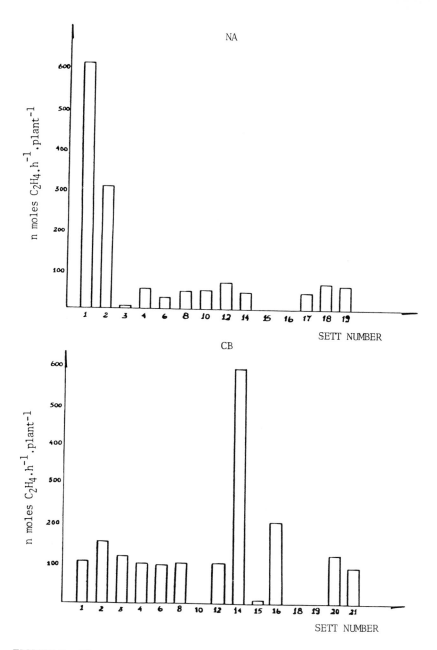

FIGURE 2. Nitrogenase activity (n mol C₂H₄/hr) of setts of varieties Na and CB
taken from the field, surface sterilized and aseptically planted in sterile vermiculite.
Numbered from the root upwards. After 3½ weeks from planting.

is supported by the fact that no acetylene reduction was obtained in nongerminated
stalks.

Dilutions from the different parts of the plant without the tops had a small number
of N_2-fixing bacteria, there being none in sett no. 18, variety CB (Table 2). Table 2
also shows that there are more bacteria in the interior of the stalks of variety NA,
since ARA was positive up to dilution 10^3 in two stalks analyzed. In this case, stalks
that contributed a larger number of N_2-fixing bacteria to the rhizosphere can be selected for future work.

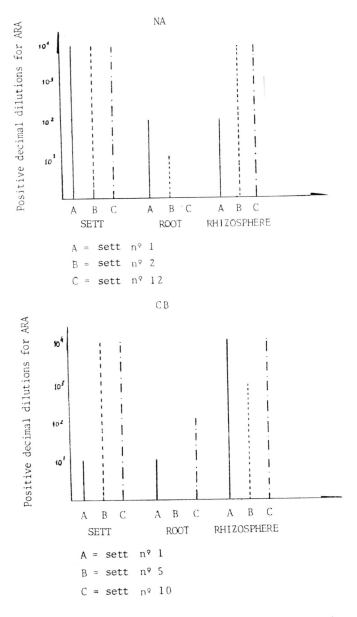

FIGURE 3.   Positive dilutions for ARA, in culture media LGP, in different plant parts of varieties Na and CB, grown under greenhouse conditions which had not shown ARA in intact systems.

C. Plants Cultivated in Nutrient Solution

Table 4 shows that there was a significant difference in ARA of pieces of roots incubated in two culture media, LG and LGY, with plants grown with and without N, which means a nutritional requirement by sugar cane N_2-fixing bacteria. There was a marked difference in ARA between root ends and parts close to the stem, and in plants cultivated without N, the bacteria showing a preference for thicker roots. This preference has already been detected in rice roots.[5]

Many experiments have demonstrated that high levels of fertilizer-N can completely inhibit nitrogenase activity in roots of *Gramineae*.[13,10] Decimal dilutions of roots of plants cultivated with different levels of N (Table 3) in the different media: LG, LGY, and LGP, showed that with 3.5 ppm N the number of ARA positive dilutions was higher for both varieties, indicating that bacteria established better in the roots when a little NO_3^- was given to the culture solution.

Sugar cane response to N-fertilizaion under normal crop conditions is frequently very small. Plants grown with 3.5 ppm NO_3^- showed high N assimilation[20] and high ARA positive dilutions, therefore it is possible that the N_2-fixing system of sugar cane requires only a small amount of N, under normal crop conditions, and which can be beneficial to both biological N_2-fixation and mineral-N assimilation (already observed in maize[13]).

D. Microbiological Studies

Pieces of roots in LG and LGY media, after 3 days incubation, were visibly different: the first had colonies 0.3 to 1.2 cm diameter, white, nontranslucent, wrinkled, and the main bacteria found was *Azotobacter*. In the second, the colonies were 0.5 to 1.5 cm diameter, yellowish, opaque color, mucous, and with many gas bubbles. Examination by light microscope revealed a large variety of bacteria: *Azotobacter, Beijerinckia, Clostridium, Bacillus,* similar to those found by Ruschel et al.[17]

E. Roots Incubated in Triphenyltetrazolium Chloride (TTC)

TTC reduction has been widely used to demonstrate dehydrogenase activity in cells and enzyme preparations.[19] Fay and Kulasoorija[9] showed that reduction of TTC is quicker in heterocysts of blue-green-algae than in vegetative cells. Although not specific, this technique is very valuable for demonstrating bacteria in the interior of plant tissue.[12,11]

Examination by light microscope, after an incubation period, of whole roots or small transverse sections of roots of plants germinated on vermiculite or nutrient solution, revealed a large number of TTC reducing bacteria populations on the cell surface of thicker roots (Plate 1, A) which included a large amount of root hairs (Plate 1, B and C). Plate 1, D shows these populations quite distinctly in the interior of the xylem, similarly to those found by Patriquin et al.[12]

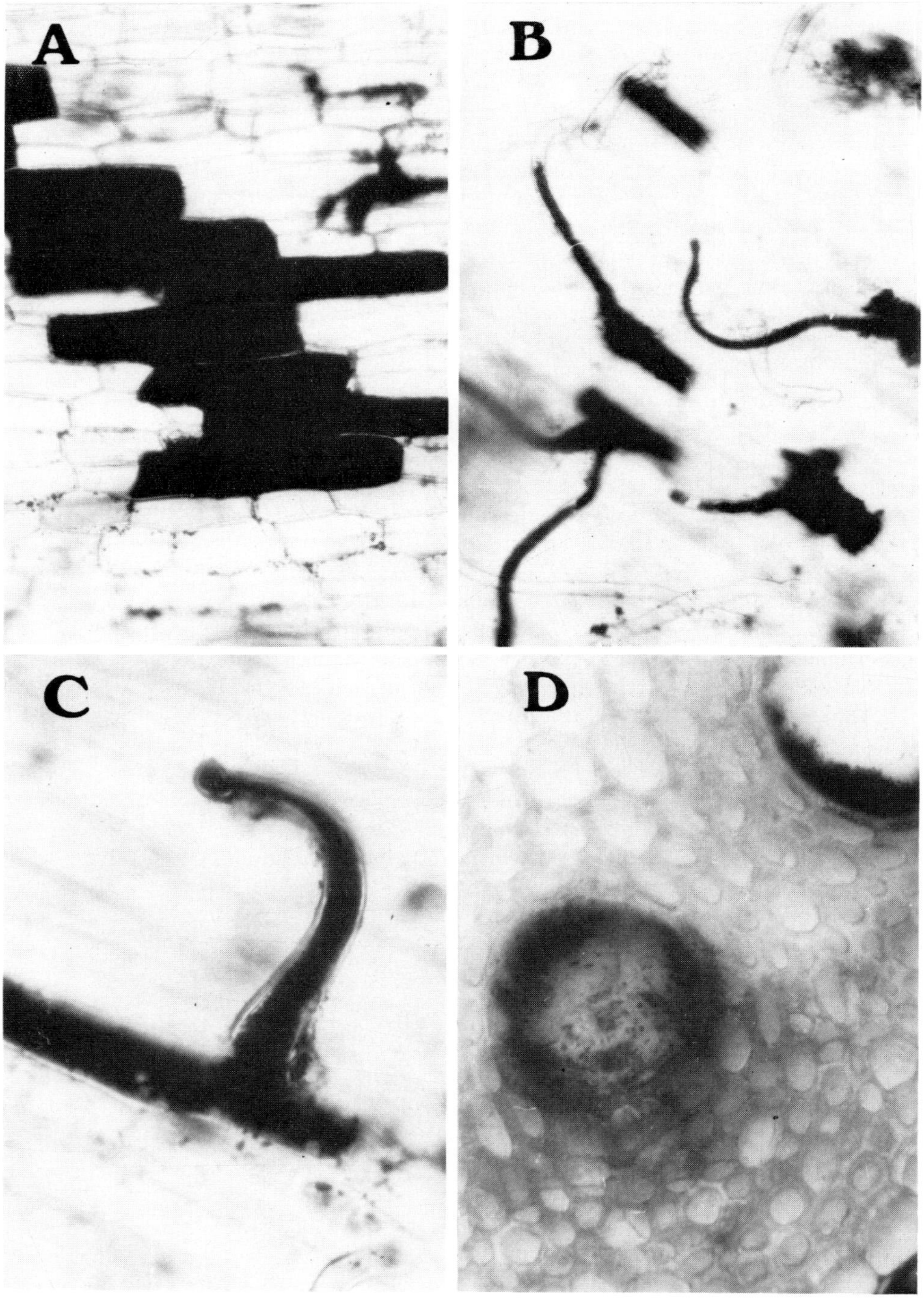

PLATE 1. A, B, and C: superficial views of root cells close to the stalk, incubated for one night in TTC, showing tetrazolium reducing bacterial colonies. A and B (Magnification × 80.) C: detail of root hair. (Magnification × 160.) D: transverse section of root, incubated for one night in TTC, showing tetrazolium reducing bacterial colonies in the interior of the xylem. (Magnification × 160.)

REFERENCES

1. **Balandreau, J., Millier, C. R., and Dommergues, Y. R.,** Diurnal variations of nitrogenase activity in the field, *Appl. Microbiol.*, 27(4), 662, 1974.
2. **Brown, M. E.,** Role of *Azotobacter paspali* in association with *Paspalum notatum, J. Appl. Bacteriol.*, 40, 341, 1976.
3. **Dart, P. J., Day, J. M., and Harris, D.,** Assay of nitrogenase activity by acetylene reduction, in *Use of Isotopes for Study of Fertilizer Utilization by Legume Crops,* Int. Atomic Energy Tech. Rep. 149, International Atomic Energy Agency, Vienna, 1972, 85.
4. **Day, J. M. and Dobereiner, J.,** Physiological aspects of N_2-fixation by a *Spirillum* from *Digitaria* roots, *Soil Biol. Biochem.*, 3, 45, 1976.
5. **Diem, G., Rougier, M., Hamad-Fares, I., Balandreau, J. P., and Dommergues, Y. R.,** Colonization of rice roots by diazotroph bacteria. Environmental role of nitrogen-fixing blue-green algae and asymbiotic bacteria, *Ecol. Bull. Stockholm;* 26, 305, 1978.
6. **Döbereiner, J.,** Influência da cana-de-acucar na população de *Beijerinckia* do solo, *Rev. Bras. Biol.,* 19, 251, 1959.
7. **Döbereiner, J. and Alvahydo, R.,** Sobre a influência da cana-de-acucar na ocorrência de *Beijerinckia* no solo. II. Influência das diversas partes do vegetal, *Rev. Bras. Biol.,* 19, 401, 1959.
8. **Döbereiner, J. and Day, J. M.,** *Dinitrogen Fixation in the Rhizosphere of Tropical Grasses,* Vol. 1, IBP Conf. Nitrogen Fixation and the Biosphere, Stewart, W. D. P., Ed., Cambridge University Press, Edinburgh, 1975, 39.
9. **Fay, P. and Kulasoorija, S. A.,** Tetrazolium reduction and nitrogenase activity in heterocystous blue-green algae, *Arch. Mikrobiol.*, 87, 341, 1972.
10. **Klucas, Y. R. and Pedersen, W.,** Nitrogen Fixation Associated with Roots of Sorghum and Wheat, Journal Series, Nebraska Agricultural Experiment Station, 1978.
11. **Patriquin, D. G. and Döbereiner, J.,** Light microscope observation of tetrazolium-reducing bacteria in the endorhizosphere of maize and other grasses in Brasil, *Can. J. Microbiol.,* 24(6), 734, 1978.
12. **Patriquin, D. G., Graciolli, L. A., and Ruschel, A. P.,** Nitrogenase Activity and Bacterial Infection of Roots and Germinated Cuttings, presented to Steenbock-Kettering Int. Symp. Nitrogen Fixation, Madison, Wis., 1978.
13. **Pereira, P. A., von Bülow, J. F. W., and Neyra, C. A.,** Atividade da nitrogenase, nitrato reductase e acumulacão de nitrogênio em milho Braquitico *Zea mays* L. (Cv. Pirañao) em dois níveis de adubacão nitrogenada, *Rev. Bras. Ci. Solo,* 2, 28, 1978.
14. **Rinaudo, G.,** Algae and bacterial non-symbiotic nitrogen fixation in paddy soils, *Plant Soil,* Special Vol., 471, 1971.
15. **Ruinen, J.,** Occurrence of *Beijerinckia* species in the ''phyllosphere'', *Nature London,* 177, 220, 1956.
16. **Ruschel, A. P.,** Aerobic and anaerobic N-fixing microorganisms in soil grown sugar cane as affected by cane plant. N-fixing inoculation and soil-addition of cane leaves, *Proc. Int. Soc. Sugar Cane Technol.*, 1978.
17. **Ruschel, A. P., Orlando Fo., J., Zambello, E., Jr., and Henis, Y.,** Aerobic and anaerobic nitrogen-fixing bacteria on sugar cane roots, *Proc. Int. Soc. Sugar Cane Technol.*, 1978.
18. **Tarrand, J. J., Krieg, N. R., and Döbereiner, J.,** A taxonomic study of the *Spirillum lipoferum* group, with descriptions of a new genus *Azospirillum* gen. nov. and *Azospirillum brasilense* sp. nov., *Can. J. Microbiol.,* 24, 967, 1978.
19. **van Fleet, D. S.,** Histochemical localization of enzymes in vascular plants, *Bot. Rev.,* 18(5), 354, 1952.
20. **Vose, P. B., Ruschel, A. P., and Victoria, R. L.,** Potential N_2-fixation by sugar cane, *Saccharum* sp., seedlings in solution culture. I. Effect of NH_4^{++} vs. NO_3^-, varieties and nitrogen level, in *Associative N_2-Fixation*, Vose, Peter B. and Ruschel, A. P., Eds., CRC Press, Boca Raton, Fla., 1980.

Chapter 14

ROLE OF *AZOTOBACTER* IN SUGAR CANE CULTURE AND THE EFFECT OF PESTICIDES ON ITS POPULATION IN SOIL

Kishan Singh, A. P. Sinha, and V. P. Agnihotri*

TABLE OF CONTENTS

*   Indian Institute of Sugar Cane Research, Lucknow, India.

I. INTRODUCTION

At least eight genera of nitrogen-fixing bacteria, namely, *Azotobacter, Bacillus, Beijerinckia, Derxia, Klebsiella, Pseudomonas, Spirillum* and *Vibrio* have been isolated from the rhizosphere of sugar cane.[9,11,14,22,28,19] Of these, species of *Azotobacter, Beijerinckia* and *Spirillum* are abundantly present in the rhizosphere of sugar cane and have been studied in depth.[9,22]

Sugar cane, a long duration crop, is attacked by a number of pathogens and insect pests[2,30] and has to compete with weeds, especially in the early stages of crop growth. For effective management of these, a number of toxicants are now in vogue. In view of increasing consumption of pesticides, it was deemed necessary to examine the effect of toxicants commonly used in sugar cane culture on *Azotobacter* a common nitrogen fixing bacteria of sugar-cane rhizosphere. The effect of *Azotobacter* on the growth of sugar cane variety Co.1148 was also investigated. The results of the two studies are presented in this paper.

II. MATERIALS AND METHODS

Composite soil samples were drawn to a depth of 15 cm at the farm of the Indian Institute of Sugar Cane Research, Lucknow (Northern India) 27.4°N, 80.9°E. It was promptly sieved and used.[1,7]

Six toxicants, namely Bavistin and Aretan (fungicides), Lindane and Phorate (insecticides), Vapam (nematicide), and Diuron (herbicide) were individually mixed, the quantity calculated on oven-dried basis (o.d.b.) weight of soil, in various concentrations, on active ingredient basis. Appropriate controls (soil without chemical) were also maintained. The soil moisture was adjusted to 60% of the water-holding-capacity (w.h.c.) and maintained at that level by adding water throughout the incubation period of 45 days. Soil samples (10 g) from treated and untreated lots were removed on the 7th, 15th, 30th, and 45th day of incubation. *Azotobacter* population was estimated by dilution plate technique on modified Jensen's® Agar:[1] sucrose 10.0 g; KH_2PO_4 1.0 g; $MgSO_4 \cdot 7H_2O$ 0.5 g; NaCl 0.5 g; $FeSO_4 \cdot 7H_2O$ 0.1 g; Na_2MS_4 0.005 g, $CaCO_3$ 1.0 g; pentachloronitrobenzene (PCNB) 50 mg, actidione 40 mg; pimaricin 35 mg; agar-agar 15 g; and distilled water 1000 mℓ. Addition of PCNB and antifungal antibiotics (pimaricin and acticione) facilitated counting of *Azotobacter* colonies by suppressing unwanted growth of actinomycetee and fungi.

Effect of *Azotobacter* on sugar cane was studied with four strains of the bacterium, which fixed more than 14 mg N/g carbon, (procured from the Indian Agricultural Research Institute, New Delhi). Sugar-cane variety Co. 1148 was used in this test. Three bud setts were steeped in *Azotobacter* culture for half an hour and placed in furrows, 90 cm apart in microplots. *Azotobacter* suspension was also sprinkled with a rose-can over the setts before closing the furrows. Urea (150 kg/ha) was applied in two split doses after 45 and 90 days of planting. For comparison a control, wherein no *Azotobacter* cultures were added, was maintained. Normal cultural and manual plant protection measures were taken during the crop growth.

III. RESULTS AND DISCUSSION

Data on chemical properties of soil and its microbial status are given in Table 1. The soil was a sandy loam having high C:N ratio. It was rich in available phosphorus (45 ppm or 90 kg/ha) with slightly alkaline pH. It contained high population of different microbes, like bacteria, actinomycetes, and fungi. The normal *Azotobacter* population was 23×10^3/g of soil at the time of starting the experiments.

Table 1
CHEMICAL AND MICROBIOLOGICAL
CHARACTERISTICS OF SOIL

Parameters	Values
a)Chemical	
CEC meg/100g	11.72
pH	7.25
Electrical conductivity at 25°C	0.106
$CaCO_3$ (%)	0.55
Total nitrogen (%)	0.022
Organic carbon (%)	0.321
C:N ratio	14.59
Available phosphorus (ppm)	45.00
K in saturated extract (ppm)	6.81
NO_3-N (ppm)	7.1
NH_4-N (ppm)	6.5
b) Microbiological	
Fungi (Martins peptone-rose-bengal agar medium)	21×10^4/g soil
50 mg PCNB Bacteria (Thornton's agar medium)	30×10^6/g soil
40 mg actidione Spore forming bacteria —	24×10^4/g soil
35 mg pimaricin Smith's agar medium	
Azotobacter — Jensen's agar medium	23×10^3/g soil
Actinomycetes (Water agar medium)	21×10^6/g soil

Table 2
EFFECT OF ARETAN AND BAVISTIN (FUNGICIDES) ON
AZOTOBACTER POPULATION ($\times 10^3$/g) IN SOIL

Aretan[a] (ppm)	Days of incubation				Bavistin[b] (ppm)	Days of incubation			
	7	115	30	45		7	15	30	45
0	26	28	24	22	0	34	37	36	33
2.5	20	22	21	23	5.0	30	33	33	30
5.0	23	23	24	25	10.0	26	29	30	29
10.0	17	19	19	18	20.0	24	25	29	24
20.0	15	18	20	21	40.0	21	24	26	25
CD at 5%	6.9	6.5	N.S.[c]	N.S.[c]	CD at 5%	7.4	8.7	6.0	5.9

[a]   Aretan = methoxy-ethyl mercury chloride — (0.6 Kg Hg/ha) dosage. The dosage used is four times higher than recommended.
[b]   Bavistin = methyl 2-benzimidazole carbamato (BASF) — (10 Kg/ha) — normal dosage.
[c]   N.S.   = nonsignificant.

All concentrations of aretan, a contact organomercurial fungicide, reduced the population of *Azotobacter* over the control (Table 2). The reduction in population was directly related to the amount of fungicide initially applied. In soils amended with 10 and 20 ppm, a significant drop in the number of *Azotobacter* propagules was noticed for the first 15 days and thereafter fluctuation in population became nonsignificant. On the other hand, with bavistin (Table 2), a broad spectrum systemic fungicide, the adverse effect of the chemical at 10, 20, and 40 ppm lasted for 45 days. Long persistence of benzimidazole fungicides in soil has been demonstrated earlier.[4,19,27]

Azotobacter number significantly decreased in vapam (250 to 1000 ppm) treated soil

Table 3
EFFECT OF VAPAM (NEMATICIDE) AND DIURON (HERBICIDE) ON
AZOTOBACTER (× 10$_3$/g) POPULATION IN SOIL

Vapam[a]	Days of incubation				Diuron[b]		Days of incubation		
(ppm)	7	15	30	45	(ppm)	7	15	30	45
0	23	25	22	21	0	27	28	25	25
125	21	27	25	20	1	24	26	24	26
250	19	24	25	25	5	22	24	26	28
500	17	21	22	32	10	19	23	22	23
1000	13	18	24	35	20	17	20	21	24
CD at 5%	6.45	N.S.[c]	N.S.[c]	N.S.[c]	CD at 5%	N.S.[c]	N.S.[c]	N.S.[c]	N.S.[c]

[a]Vapam = 32.7% sodium methyl dithiocarbamate — (250 kg to 1000 kg/ha) — normal dosage.
[b]Diuron = 3 (3,4-dichlorophenyl)-1, 1-dimethylurea. 1. (Normal dosage 1 kg/ha). 2. Used mostly as preemergence.
[c]N.S.   = nonsignificant.

Table 4
EFFECT OF LINDANE AND PHORATE (INSECTICIDES) ON *AZOTOBACTER*
(× 10^3/g) POPULATION IN SOIL

Lindane[a]	Days of incubation				Phorate[b]		Days of incubation		
(ppm)	7	15	30	45	(ppm)	7	15	30	45
0	24	27	28	26	0	27	27	28	26
1	25	26	29	26	1.5	26	27	27	24
5	23	25	27	24	7.5	28	29	28	27
10	21	23	24	21	15.0	22	25	26	22
20	21	23	24	23	30.0	22	23	24	26
CD at 5%	N.S.[c]	N.S.[c]	N.S.[c]	N.S.[c]	CD at 5%	N.S.[c]	N.S.[c]	N.S.[c]	N.S.[c]

[a]   Lindane = 1,2,3,4,5,6 hexa-chloro cyclohexane (gamma isomer) normal dosage 1 kg/ha.
[b]   Phorate = 0, 0-diethyl-5 (ethylthiomethyl) phosphorodithioate (normal dosage 3 kg/ha).
[c]   N.S.   = Nonsignificant.

by the seventh day and thereafter fluctuation in population became nonsignificant (Table 3). On the 45th day, significant increase in their population was discernible in treatments with 500 and 1000 ppm vapam. This enhancement in population could be due to the reduction in competition from slower growing soil microbes, efficient utilization by *Azotobacter* isolates of degradation products from dead fauna and flora due to vapam, a nonspecific biocide. Similar stimulatory effect after several weeks of applying this chemical was reported also by Naumann.[26] Diuron, a preemergence herbicide, did not significantly alter the population of the bacterium as compared to the control (Table 3) except with 7 days of incubation at higher dosages. Goguadze[17] demonstrated that diuron not only inhibited the population of the bacterium but also adversely affected nitrogen fixation. The difference in results could be due to differences in strains of the bacterium, dosages and/or soil types used in these studies.

Common soil insecticides phorate and lindane were neither stimulatory nor inhibitory to soil *Azotobacter* (Table 4). Chiaro[8] reported that chlordane (5 to 10 kg/ha) temporarily depressed the activity of *Azotobacter*. Gaur and Bardiya[16] found that at higher concentrations, lindane reduced the population by the 14th day, whereas dieldrin did not bring about any significant change in the population. Mahmoud et al.,[23] on the other hand, reported that *Azotobacter* numbers were increased in lindane and dieldrin amended soils.

Table 5
EFFECT OF *AZOTOBACTER* **CULTURES ON GERMINATION, TILLERING, AND YIELD OF 12-MONTH CROP OF SUGAR CANE CULTIVAR CO.1148**

Treatment	% germination	Tiller/ plant	kg/yield/ plot (m²)	% increase over control
Azotobacter-I	31.6	4.8	371	8.1
Azotobacter-II	30.4	5.3	392	13.0
Azotobacter-III	29.4	5.2	370	7.8
Azotobacter-IV	30.0	5.5	379	10.0
Azotobacter mixture	27.9	5.7	399	14.5
Control	34.0	5.3	341	
CD at 5%			N.S.[a]	

[a] N.S. = nonsignificant

Data in Table 5 show that *Azotobacter* cultures when applied to sugar-cane variety Co.1148 increased the yield of the 12-month crop from 7.6 to 14.5% over the control. Since the yield increase was not significant, it is hard to evaluate the ecological significance of nitrogen fixation by nonsymbiotic *Azotobacter* cultures with the present data. Some investigators have claimed that nitrogen enrichment by *Azotobacter* is almost negligible.[25] Conversely, Döbereiner[10] felt that under tropical climate and in the rhizosphere of plants, nonsymbiotic nitrogen fixation by *Azotobacter* needed in depth studies. The increase in yield has also been ascribed to the secretion of growth promoting substances, e.g., auxin, gibbrellin, cytokinin by this bacterium.[3,5,6,32] Presence of these substances in the rhizosphere region is known to stimulate plant metabolism which, in turn, increases uptake of nitrogen, phosphorus, and other elements.[15] From the present study it is not possible to identify the factor, repsonsible for increasing yield of sugar-cane variety Co.1148.

In conclusion, it may be said that the toxicants used at the dosages normally recommended for commercial sugar-cane culture did not produce any adverse effect on the population of *Azotobacter* after 15 days. Though statistically nonsignificant there was a slight drop in population after the first 7 days in case of the fungicides and the nematicide. It would therefore be safe to use the agricultural chemicals mentioned here, without any adverse effects on *Azotobacter* population of the soil. The positive response, though low, opens up new avenues of research in sugar-cane production technology and deserves to be more extensively and systematically investigated, keeping in view the effect(s) of edaphic factors on build up of the *Azotobacter* population, especially method of inoculation and time of application.

REFERENCES

1. **Agnihotri, V. P.,** Persistence of captan and its effect on microflora, respiration and nitrification of a forest nursery soil, *Can. J. Microbiol.*, 17, 377, 1971.
2. **Agnihotri, V. P. and Singh, Kishan,** Seed piece transmissible sugar cane diseases and their control measures, *Sugar News*, 9, 90, 1977.

3. **Barea, J. M., Navarro, E., and Montoya, E.,** Production of plant growth regulators by rhizosphere phosphate dissolving bacteria, *J. Appl. Bacteriol.,* 40, 129, 1976.

4. **Baude, F. J., Pease, H. L., and Holt, R. F.,** Fate of benomyl on field soil and turf, *J. Agric. Food. Chem.,* 22, 413, 1974.

5. **Brown, M. E.,** Plant growth substances produced by microorganisms of soil and rhizosphere, *J. Appl. Bacteriol.,* 35, 443, 1972.

6. **Brown, M. E. and Walkar, N.,** Indole-3-acetic acid produced by *Azotobacter chroococcum, Plant Soil,* 32, 250, 1970.

7. **Chandra, P. and Bollen, W. B.,** Effect of nabam and mylone on nitrification, soil respiration and microbial numbers in four Oregon soils, *Soil Sci.,* 92, 387, 1961.

8. **Chiaro, G.de.,** The effect of chlordane on the soil microflora, *Agric. Ital.,* 53, 347, 1953.

9. **Döbereiner, J.,** Nitrogen fixing bacteria of the genus *Beijerinckia* Berx. in the rhizosphere of sugar cane, *Plant Soil,* 14, 211, 1961.

10. **Döbereiner, J.,** Non-symbiotic nitrogen fixation in tropical soils, *Pesqui. Agropecu. Bras.,* 3, 1, 1968.

11. **Döbereiner, J. and Campelo, A. B.,** Non-symbiotic nitrogen fixing bacteria in tropical soils, *Plant Soil,* Special volume, 457, 1971.

12. **Döbereiner, J. and Day, J. M.,** Nitrogen fixation in the rhizosphere of tropical grasses, in *Nitrogen Fixation by Free Living Microorganisms,* Stewart, W. D. P., Ed., Cambridge University Press, London, 1975, 39.

13. **Döbereiner, J., Day, J. M., and Dart, P. J.,** Nitrogenase activity in the rhizosphere of sugar cane and some other tropical grasses, *Plant Soil,* 37, 191, 1972a.

14. **Döbereiner, J., Day, J. M., and Dart, P. J.,** Nitrogenase activity and oxygen sensitivity of the *Paspalum notatum-Azotobacter paspali* association, *J. Gen. Microbiol.,* 71, 103, 1972b.

15. **Borosnikii, L. M.,** Some questions on the use of bacterial fertilizers (Eng. trans.), *Mikrobiologiya,* 31, 738, 1962.

16. **Gaur, A. C. and Bardiya, M. C.,** The effect of lindane and dieldrin on soil microbial population, *Indian J. Microbiol.,* 10, 33, 1970.

17. **Goguadze, V. D.,** Effect of herbicides on the development of *Azotobacter* in some soils of Western Georgia, *Agrokhimiya,* 3, 99, 1968.

18. **Hiltner, L.,** Uber neuere Erfahrungen and Problems auf dem. Gebiet der Bodenbakteriologie und unter besondeder Berucksicktigung der Grun dungung und Brache, *Arb. Dtsch. Landw. Ges.,* 98, 59, 1904.

19. **Hine, R. B., Johnson, D. L., and Wenger, C. J.,** The persistency of the benzimidazole fungicides and their fungistatic activity against *Phymatotrichum omnivorum, Phytopathology,* 59, 798, 1969.

20. **Jensen, H. L.,** Notes on the biology of *Azotobacter, Proc. Soc. Appl. Bacteriol.,* 14, 89, 1951.

21. **Kamal and Singh, N. P.,** On microfungi from root region of ten sugar cane varieties, *Indian Phytopathol.,* 37, 347, 1974.

22. **Khanna, S.,** Studies on the Non-Symbiotic Nitrogen Fixing Bacteria from the Rhizosphere of Sugar Cane (*Saccharum* sp.), M.Sc. thesis, Division of Microbiology, Indian Agricultural Research Institute, New Delhi, 1978.

23. **Mahmoud, S. A. Z., Salim, K. C. and Elmokatem, M. T.,** Effect of dieldrin and lindane on soil microorganisms, *Zentrabl. Bakteriol. Parasitkde. Infectionskr. Hyg. Abt. 2,* 125, 134, 1970.

24. **Majumdar, S. B. and Bhide, V. P.,** Influence of root exudates of sugar cane on rhizosphere fungi, *J. Univ. Poona. Sci. Technol.,* 40, 91, 1971.

25. **Mishustin, E. N. and Shilnikova, V. K.,** The Biological Fixation of Atmospheric Nitrogen by Free Living Bacteria, in *Soil Biology,* UNESCO Publ., New York, 1969, 62.

26. **Naumann, K.,** The effect of some environmental factors on the reaction of the soil microflora to plant protection agents (in German), *Zentrabl. Bakterio. Parasitkde. Infectionskr. Hyg. Abt. 2,* 127, 379, 1972.

27. **Raynal, C. and Farrari, F.,** Persistence of soil incorporated with benomyl and its effect on soil fungi, *Phytriatrie-Phytopharmacie,* 22, 259, 1973.

28. **Ruschel, A. P., Filho, J. O., Zambello, E., Jr., and Henis, Y.,** Aerobic and anerobic nitrogen fixing bacteria on sugar cane roots, *Int. Soc. Sugar Cane Technol.,* 1923, 1978.

29. **Ruschel, A. P., Henis, Y., and Salati, E.,** Nitrogen-15 tracing of nitrogen fixation with soil grown sugar cane seedlings, *Soil Biochem.,* 5, 181, 1975.

30. **Singh, K., Agnihotri, V. P., and Avasthy, P. N.,** Diseases and pests of sugar cane, Proc. DSTA. 1, 1973.

31. **Srinivasan, K. V.,** Fungi of the rhizosphere of sugar cane and allied plants. I. Hyalostachybotrys gen. nov., *J. Indian Bot. Soc.,* 37, 334, 1958.

32. **Vancura, V. and Macura, J.,** Indole derivatives in *Azotobacter* cultures, *Folia Microbiol. Prague,* 5, 393, 1960.

Chapter 15

SEASONAL VARIATION IN THE MICROBIAL POPULATIONS OF SUGAR-CANE PLANTS*

J. M. T. Ferro Costa and A. P. Ruschel**

TABLE OF CONTENTS

*   Translator:Diva Athié.

** Centro de Energia Nuclear na Agricultura, Piracicaba, S. P., Brasil.

I. INTRODUCTION

According to Patriquin et al.,[2] bacterial populations are established both in the internal part of the cortex and the external part of the stele. They noted, however, that bacterial populations in roots of germinated stalks occur mainly in the epidermal intercellular oblong spaces, close to where the roots emerge from the cane nodes. From these data these authors believe that there is a bidirectional movement of the bacteria through these intercellular spaces in germinated sugar-cane stalks.

Considering this possible bidirectional movement of the bacteria, it appeared desirable to check nitrogenase activity distribution in a series of germinated stems of the same plant before and after germination. Would all germinated stalks from the stem sequence have the same nitrogenase activity? If not, would the activity of the germinated stalks decrease with distance from soil to root? How would seasonal factors influence bacterial migration and the population of the rhizosphere in the sequence of germinated stalks?

The present work aims at helping to answer such basic questions, which are of interest to research on biological N$_2$-fixation in sugar cane.

II. MATERIAL AND METHODS

Various experiments were carried out with the objective of studying nitrogenase activity in germinated and nongerminated stems. The experiments with germinated stems were carried out at three different seasons: spring, summer, and autumn. Varieties NA56-79 and CB46-47 were used in the first two experiments. The only variety used in the autumn experiment was NA58-79.

The stalks for the experiments were obtained from the field and used at most 2 days after cutting. Before the experiment, each stalk was cut, numbered in relation to distance from the root and submitted to a superficial sterilization process with commercial sodium hypochlorite (Q-Boa) at 10% dilution and washed in water and sterile buffer. In the last two experiments nonsterile stalks were also planted, under the same aseptic conditions, for control and comparison.

Nitrogenase activity in nongerminated stalks of mature plants was studied by grinding them in the blender under aseptic conditions, in a sterile buffer solution 0.05 M pH 7.0 (10 × v/w) for 1 to 2 min. Decimal dilution was made in culture medium tubes, and the 10^3 and 10^5 were chosen and inoculated in the following culture media: LG, LG without CaCO$_3$, LGY and LGY incubated in anaerobises by reaction of pyrogalic acid + sodium carbonate,[5] LG-Malate,[1] GL-Malate-Glucose.[2] These five different culture media were used to make it possible to grow all kinds of asymbiotically N$_2$-fixing bacteria in a medium corresponding to nutritional requirements, growing conditions and the characteristics inherent to each group.

When growth started in the tubes, they were incubated with 0.01 at C$_2$H$_2$ (1%) at 28°C. Nitrogenase activity was determined 1 hr later with a Beckman® gas chromatograph, model GC65, using an H$_2$ flame ionization detector at 175°C and a glass column with internal diameter 1/8 in. × 1.60 cm, containing Porapak N, 8-100 mesh, at 110°C. N$_2$ was the carrier-gas used at a flow rate of 40nℓ/min.

In the germiantion experiments the sterile stalks were planted in pots containing vermiculite and sterile sand (v/v), maintained in a glass-house and irrigated with a modified Hoagland solution, deficient in nitrogen and diluted to 1/5.

Nitrogenase activity (n mol C$_2$H$_2$/hr) was determined in plantlets about 15 days after germination, in intact and disturbed systems: top parts, stem, roots, and rhizosphere soil. Plants were incubated with 0.1 atm C$_2$H$_2$ (10%) at room temperature and measurements taken at 1, 2, and 4 hr.

One plant was incubated without C_2H_2 as a control for endogenous ethylene, as well as a pot with vermiculite as a control for the blank calculation.

III. DISCUSSION AND CONCLUSIONS

In the experiments with nongerminated stalks, several inoculated tubes of both dilutions 10^3 and 10^5, with evidence of microbial growth, showed no nitrogenase activity — nmol C_2H_4/hr under the incubation conditions used, acetylene reduction activity (ARA) being low in the positive tubes (Table 1). It is believed that some unknown factor inhibits ARA of the N_2-fixing bacteria in the interior of the stalk, keeping them latent until germiation.[2] The tubes inoculated with the ground mixture from distal and proximal stalks showed greater nitrogenase activity. It is possible that colonization and infection by N_2-fixing bacteria of sugar cane roots[2] as well as populations of the phyllosphere of *Gramineae*[3,4] can act as an initial inoculum for establishment of the infection and subsequent population of the stalks.

The greatest nitrogenase activity and bacterial growth was observed in the LG-malate-glucose medium, with all stalks, indicating that the majority of the heterotrophic N_2-fixing microorganisms present in the interior of the cane are microaerophilic and capable of fermenting glucose and/or malate (Table 1).

Microorganisms grown both in LG and LG-malate media showed nitrogenase activity, although little, in all stalks. The low activity of the aerobic bacteria (LG medium) seems to indicate a smaller aerobic population, the same being the case with microorganisms grown in LGY, LGY-anaerobic, and LG-CaCO₃ media, especially from the top of the stalks.

The various plantlets obtained from germination of the stalk series showed quite varied nitrogenase activity (Figures 1, 2, 3, and 4). Germinated stalks do not all show the same potential for nitrogenase activity, the greatest activity apparently being concentrated in the distal stalks.

There is an apparent seasonal influence on the distribution of N_2-fixing bacteria in the different stalks from the same plant. Spring distribution of nitrogenase activity within the stalk is very irregular, in summer it is relatively uniform, with the level falling abruptly in autumn.

By comparison of Figures 2, 3 and 4 it can be noted that nitrogen-fixing populations in the interior of the stalks of superficially sterilized treatments are similar to the total stalk (nonsterilized treatments). However, during autumn the activity of the microorganisms in the interior of the stalk drastically decreases (Figure 4), which is not observed when analyzing plants from nonsterilized stalks.

Nitrogenase activity of disturbed systems was higher in the stalks and rhizosphere soil than in the roots. The high nitrogenase activity of the soil suggests a movement of the bacteria from the interior of the stalk to the rhizosphere soil after germination (Tables 2 and 3). The highest activity was observed in summer, even in the top parts.

In summary: apparently the majority of heterotrophic N_2-fixing microorganisms in sugar-cane stalks are microaerophilic; not all germinated stalks have the same potential for nitrogen fixation, this tending to be greatest in the upper stem pieces; distribution of nitrogenase activity within the stalk is irregular in spring, relatively uniform in summer, and declining abruptly to a constant low level in autumn.

Table 1

NITROGENASE ACTIVITY (ARA) — NMOL C_2H_4/HR IN MACERATED STALK DILUTIONS (10^3 AND 10^5) OF THE PROXIMAL, INTERMEDIATE AND DISTAL INTERNODES OF TWO SUGAR-CANE VARIETIES

Varieties	Stalks	Dilution	Microaerophilic (LG mal. gl.)	Aerobic (LG)	Microaerophilic (LG mal.)	Facultative anaerobe (LGY)	Facultative anaerobe (LG)	Acidophilic microorganisms (LG-CaCO₃)
CB 46-47	Proximal	10^3	0	2.85	0	0	0	0
		10^5	0	2.71	5.11	0	0	0
	Intermediate	10^3	5.41	0	1.48	0	0	35.77
		10^5	1.20	1.27	1.41	0	0	0
	Distal	10^3	3631.4	5.24	4.5	73.77	0	21.55
		10^5	353.3	1.95	0	27.63	0	38.94
NA 56-79	Proximal	10^3	5358.51	4.28	3.67	6.96	0	2.91
		10^5	303.09	0	18.58	0	0	3.05
	Intermediate	10^3	7.35	0	1.36	0	0	0
		10^5	6.73	2.43	3.06	0	0	0
	Distal	10^3	6.78	4.13	0	0	7.95	0
		10^5	7.95	2.74	0	0	2.22	0

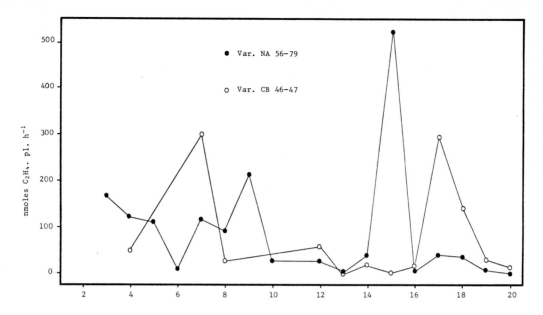

FIGURE 1.   Nitrogenase activity (ARA) — nmoles C_2H_4/hr of plantlets germinated from stalks numbered in relation to soil. Spring 1978. Averages of two replicates.

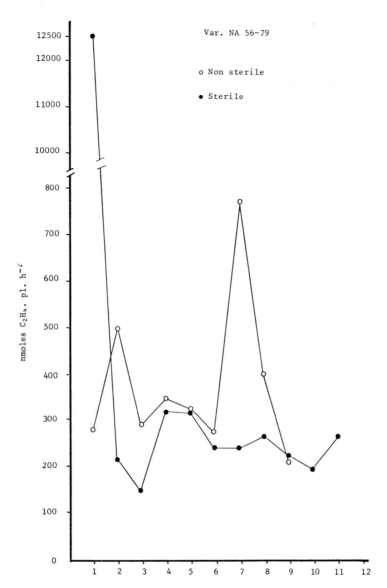

FIGURE 2.   Nitrogenase activity (ARA) — nmoles C_2H_4/hr of plantlets germinated from stalks numbered in relation to soil. Summer 1979. Averages of two replicates.

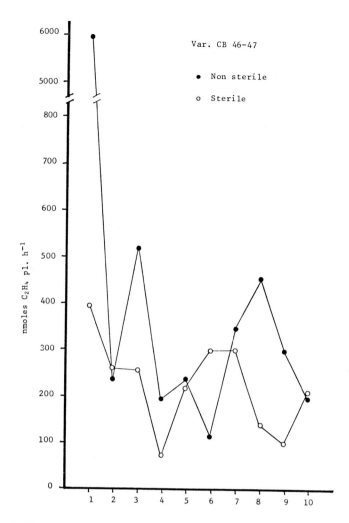

FIGURE 3. Nitrogenase activity (ARA) — nmol C_2H_4/hr of plantlets germinated from stalks numbered in relation to soil. Summer 1979. Averages of two replicates.

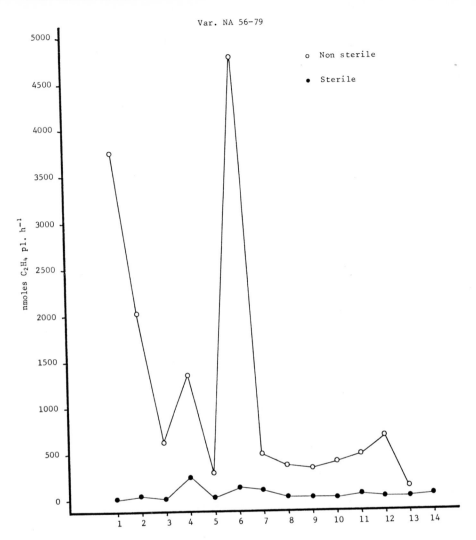

FIGURE 4. Nitrogenase activity (ARA) — nmol C_2H_4/hr of plantlets germinated from stalks numbered in relation to soil. Fall 1979. Averages of two replicates.

Table 2
NITROGENASE ACTIVITY (ARA) — NMOL C_2H_4/ HR OF SUGAR-CANE PLANTLETS OBTAINED FROM STERILE CUTTINGS, VAR. NA 56-79 AND CB 46-47, IN DISTURBED SYSTEM: LEAF, STALK ROOT, RHIZOSPHERE SOIL, 15 DAYS AFTER GERMINATION; SPRING 1978; AVERAGE OF TWO REPLICATES

Varieties	Germinated stalk	Leaf	Stalk	Root	Rhizosphere soil
	S-3	—	12.73	0	2.96
	S-4	—	21.40	2.77	65.67
	S-5	—	20.06	21.30	30.80
	S-6	3.24	8.33	3.64	3.21
	S-7	—	26.64	—	5.74
	S-8	4.90	15.22	4.19	4.72
	S-9	—	33.74	—	5.27
NA 56-79	S-10	—	22.42	7.96	283.69
	S-12	7.22	5.09	1.99	2.50
	S-14	7.84	12.55	3.01	3.58
	S-15	—	287.23	—	509.73
	S-17	13.08	14.40	4.38	9.50
	S-18	3.42	89.71	5.80	11.80
	S-20	—	22.68	3.80	4.16
	S-4	0	0	7.10	1.63
	S-7	—	131.18	—	6.40
	S-12	—	24.53	—	2.25
CB 46-47	S-14	0	0	0	0
	S-17	—	56.95	—	10.78
	S-18	0	7.68	2.11	0
	S-19	0	41.91	0	0
	S-20	0	4.80	0	0

Table 3
NITROGENASE ACTIVITY (ARA) — NMOL C_2H_4/HR OF SUGAR-CANE PLANTLETS, OBTAINED FROM STERILE CUTTINGS, VAR. NA 56-79, IN DISTURBED SYSTEM: LEAF, STALK, ROOT, RHIZOSPHERE SOIL, SUMMER 1979, AVERAGE OF TWO REPLICATES

Germinated stalk	Leaf	Stalk	Root	Rhizosphere soil
S-1	—	6.193,80	—	6.274,15
S-2	—	495,51	379,63	371,63
S-3	—	507,22	376,65	160,71
S-4	341,50	475,86	475,42	488,81
S-5	351,17	392,83	369,02	601,15
S-6	371,07	696,57	345,03	494,76
S-7	624,96	976,50	442,68	436,17
S-8	572,21	572,88	410,13	3.521,91
S-9	14,88	19,53	0	0
S-10	0	87,88	0	0

REFERENCES

1. **Dobereiner, J. and Day, J. M.**, Associative symbioses in tropical grasses — characterization of microorganisms and nitrogen fixing sites, in *Proc. 1st In. Symp. N$_2$-Fixation,* Newton, W. E. and Nyman, C., Eds., Washington State University Press, Pullman, 1976, 518.
2. **Patriquin, D. G., Ruschel, A. P., and Gracioli, L.**, Nitrogenase Activity and Bacterial Infection of Roots and Germinated Cuttings, Proc. of Steenbock-Kettering Int. Symp. on Nitrogen Fixation, Madison, Wis., 1978, in press.
3. **Ruinen, J.**, Nitrogen fixation in the phyllosphere, *Plant Soil,* 22, 375, 1965.
4. **Ruinen, J.**, The grass sheath, a habitat for nitrogen fixing microorganisms, *Plant Soil,* 33, 661, 1970.
5. **Veldkramp, H.**, Enrichment cultures of prokaryotic organisms, in *Methods in Microbiology,* Norris, J. R. and Ribbons, D. W., Eds., Academic Press, New York, 1970, Chap. 5.

Chapter 16

POTENTIAL N$_2$-FIXATION BY SUGAR CANE, *SACCHARUM* SP. IN SOLUTION CULTURE
I. EFFECT OF NH$^+_4$ VS. NO$^-_3$, VARIETY AND NITROGEN LEVEL

P. B. Vose*, Alaides P. Ruschel, R. L. Victoria, and E. Matsui**

TABLE OF CONTENTS

*   International Atomic Energy Agency/United Nations Development Program, Project BRA 78/006.
** Centro de Energia Nuclear na Agricultura, Piracicaba, S. P., Brazil.

I. INTRODUCTION

Surface sterilized dormant "seed pieces" (pieces of stalk with a nodal bud) of sugar cane do not show any nitrogenase activity, but if allowed to "germinate" and develop roots, then strong nitrogenase activity is shown. Direct evidence for N_2-fixation in association with sugar-cane seedlings was established by $^{15}N_2$ experiments carried out on seedlings in disturbed and nondisturbed soil-systems in a closed atmosphere with artificial light.[3,4]

Work by Arias et al.[1] demonstrated the presence of nitrogen fixing bacteria in sugar-cane roots, while Patriquin et al.[2] showed the presence of putative fixing bacteria in the stalks, especially the nodes of sugar cane and related population density to acetylene reduction activity (ARA). Moreover, elliptical ruptures were demonstrated at the base of emerging roots, around which bacteria were especially dense.

Ruschel[7] later found that if surface sterilized seed pieces were placed in sterile vermiculite and allowed to "germinate" under sterile conditions, then the vermiculite, following careful removal of the seed piece, showed high nitrogenase activity. This strongly suggested that nitrogen fixing bacteria had passed out from developing roots and were capable of fixing N_2 external to the root.

The question therefore arose as to whether the system was capable of fixing nitrogen without the mechanical and diverse biochemical support of a solid medium, and moreover whether sufficient bacteria are initially present in a "seed piece" to support N_2-fixation without further adventitious or specific inoculation. The following experiments were designed to answer these questions.

II. METHODS

Seed pieces of mature stalk, 4 cm long and containing a node with nodal bud were selected at random from two sugar-cane varieties NA56-79 and CB41-76. The seed pieces were surface sterilized with 10% hypochlorite and suspended by nylon thread from pieces of stick laid across 3 cm diameter holes in Masonite ® covers, on 18 liter plastic buckets containing continuously aerated culture solution. In the main experiment four plants were grown in each container, with three replications of each variety and N-level.

The nutrient solution consisted of 1.5×10^{-3} M KH_2PO_4, 2×10^{-3} M $MgSO_4 \cdot 7H_2O$, 1.5×10^{-3} M $CaCl_2$, 1×10^{-3} M Na_2SiO_3 (this was too much, 2.5×10^{-4} M would have been adequate), 1×10^{-4} M FeEDTA, plus complete A to Z micronutrients including Co. Nitrate-N was given as KNO_3 at rates of 3.5, 7.0, 14.0 and 28.0 ppm according to the experiment and ammonium-N as $(NH_4)_2SO_4$ at rates of 7.0 and 28.0 ppm in the preliminary experiment. Distilled water was used throughout. In the main experiment the KNO_3 had 4.04 atom % ^{15}N. The solutions were not changed during the course of the experiment, distilled water being added as necessary.

The fresh weight of all seed pieces was determined and a random sample dried for determination of percent moisture, to permit calculation of the dry weight of those used in the experiment, and for estimation of initial N-content.

The main experiment was harvested at 60 days, and the tops and "seed piece" stalks dried and weighed separately, but the pots were combined for determination of root weight, N contents and $^{15}N/^{14}N$ ratios. Determination of N and ^{15}N was carried out by methods previously described.[4]

Natural $^{15}N/^{14}N$ variation of unlabeled material was determined by "double collector" technique, labeled samples by normal "single collector" methods.

III. RESULTS AND DISCUSSION

A. Preliminary Experiment

A preliminary, largely qualitative, experiment was carried out with the variety CB41-76, in which four seed pieces per treatment were germinated in culture solution with 0, 14, and 28 ppm N, given in both NO_3^- and NH_4^+ forms. Seed pieces on zero-N and NO_3^--nitrogen germinated readily and it was noticeable that the shoot of the zero-N plants developed initially with very pale yellowish green leaves, typical of nitrogen deficiency, but after a period of about 15 to 20 days they gradually assumed a more green appearance.

Seed pieces with 7 ppm NH_4^+-nitrogen germinated more slowly, but when established grew well. With 28 ppm NH_4^+-N buds failed to develop and dried up, indicating that the NH_4^+ was toxic. However, where a bud had developed a shoot precociously, the shoot responded remarkably to the high level of NH_4^+ and grew very rapidly, suggesting that NH_4^+ is particularly toxic at the early bud stage. "Ammonium-plants" were a darker shade of green than either zero-N or "nitrate-plants".

The plants were harvested at 30 days and the plants on 7 ppm NH_4^+ had dry, weight yield of tops about 10% greater than the NO_3^--plants, and with as much as two and a half times the amount of total-N. Both total yield and N-content of the zero-N plants were much lower than any other treatment, being about half that of 7 ppm NO_3^--N plants.

Samples of roots were taken for estimation of nitrogenase activity (ARA). Roots from zero-N showed activity ($\simeq 109$ nmol C_2H_4/g/hr) within the first 30 min, but after 60 min roots from 7 ppm NO_3^- and zero-N showed about the same activity ($\simeq 75$ nmol C_2H_4/g/hr) while roots from 7 ppm NH_4^+-N showed only about 25% of this activity after 2 hr, which is extremely low. Roots from 28 ppm NH_4^+-N showed no activity, even after 3 hr.

No attempt was made to estimate possible N_2-fixation in this experiment, but it was clear from the ARA results that the NH_4^+ ion even at low levels was highly detrimental to root nitrogenase activity, and consequently only NO_3^- was used in the main experiment.

B. Main Experiment

In this experiment plants of NA56-79 and CB41-76 varieties were germinated and grown in $^{15}NO_3$ culture solution with zero-N, 3.5 and 14.0 ppm NO_3 nitrogen levels, and were harvested after 60 days growth. The yield and nitrogen contents are given in Table 1. There was little difference in yield of tops and roots between 0 and 3.5 ppm N levels, but an obvious increase with 14 ppm N, this being especially noticeable in the case of roots. This marked increase in root weight was accompanied by a decrease in percent N content of the roots, although there was a substantial increase in total N (Table 2). Increase in both percent N and total N in the tops was marked at the 14 ppm N level.

The "seed piece" stalk lost weight during the experiment, this being noticeably greater with the variety CB. The percent N content also declined substantially, and calculation of the total N content (Table 2) showed that the stalks lost about 50% of their nitrogen over 60 days. It is possibly surprising that this loss was not greater considering the low N-level of the culture solution. At 3.5 and 14.0 ppm N-levels the proportion lost was much greater in CB than NA.

Table 2 provides a breakdown of the total-N in tops, roots, and stalks, the summation of which should permit the calculation of a nitrogen balance for the system, taking into account the original N-content of the seed stalk, and where appropriate the nitrogen supplied by the culture solution. This shows that where no NO_3^- nitrogen was added

Table 1
YIELD AND NITROGEN CONTENT OF SUGAR-CANE VARIETIES NA56-79 AND CB41-76 ON ZERO AND TWO LEVELS OF NO$_3$ AND HARVESTED AT 60 DAYS

	Treatment					
	Zero N		3.5 ppm N		14 ppm N	
N level variety	NA	CB	NA	CB	NA	CB
Tops						
Dry Wt., g/plant	1.86	3.15	2.59	3.02	4.54	6.82
N-content, %	0.46	0.48	0.47	0.45	0.56	0.57
Roots						
Dry Wt., g/plant	1.68	2.19	1.90	1.91	5.35	5.13
N-content, %	0.83	0.75	0.61	0.48	0.54	0.46
Stalk ("Seed Piece")						
Initial calc. Dry Wt., g/plant	6.31	8.49	6.06	8.23	5.74	9.15
Final Dry Wt., g/plant	5.44	6.54	5.34	4.78	5.47	6.79
∴ Loss in Dry Wt., g/plant	0.87	1.95	0.72	3.45	0.27	2.36
∴ % loss in Dry Wt.	13.83	23.00	11.85	41.90	4.75	25.80
Initial N-content, %	0.46	0.47	0.46	0.47	0.46	0.47
Final N-content, %	0.25	0.26	0.29	0.28	0.32	0.36
Number of surviving plants	11	12	10	9	12	8

Note: Sig. diffs. ($P = 0.05$) for Dry Wt. tops = 0.27, for Dry Wt. roots = 0.23, for initial Dry Wt. stalks = 1.14, for final Dry Wt. stalks = 0.72, for % N tops = 0.014, for % N roots = 0.02, for final % N stalks = 0.003.

Table 2
NITROGEN BALANCE OF THE EXPERIMENT, CALCULATED FOR VARIETIES AND NITROGEN TREATMENT FROM TABLE 1

	Treatment					
	Zero N		3.5 ppm N		14 ppm N	
N level variety	NA	CB	NA	CB	NA	CB
Component						
Total N in tops, mg/plant	8.58	15.11	12.07	13.59	25.44	38.87
Total N in roots, mg/plant	13.93	16.45	11.58	9.17	28.88	23.59
Total N in "seed piece", mg/plant	13.60	17.01	15.49	13.38	17.51	24.44
∴ Total N in whole plant, mg/plant	36.11	48.57	39.15	36.14	71.83	86.90
Original N content of "seed piece", mg/plant	29.06	39.94	27.87	38.68	26.42	43.01
∴ Total N — Original N content, mg/plant	+ 7.05	+ 8.63	+ 11.28	−2.54	+ 45.41	+ 43.89
Nitrogen available from culture solution, assuming all used, mg/plant	0.0	0.0	18.90	21.00	63.00	94.50
∴ apparent increase in N in the system, mg/plant	+ 7.05	+ 8.63	calculation not possible			
and, expressed as % of total plant N	19.5%	17.7%				

Note: Sig. diffs. ($P = 0.05$) for total N tops = 1.68, total N roots = 1.44, total N "seed piece" = 0.57

to the culture solution there was a net gain of nitrogen in the system during the growing period, with the clear implication that there must have been an increase in nitrogen in the system of both NA and CB varieties due to N_2-fixation. Where NO_3 was given in the culture solution this approach was not possible, because the total amount of N that was available exceeded the increase in total-N content of the plants. Moreover, the amount of NO_3 that was available to each plant varied, because not all "seed" stalks produced surviving plants. However, the isotope data permit calculation of a balance from the analyses of atom $\% \, ^{15}N$ in the plant parts given in Table 3.

As a consequence of the isotope dilution principle, the proportion of N in any plant part derived from the nutrient solution will be given as:

$$\frac{\% \text{ N derived from}}{\text{nutrient solution}} = \frac{\text{atom } \%^{15}N \text{ excess, plant part}}{\text{atom } \%^{15}N \text{ excess, nutrient solution}} \times 100$$

$$\begin{array}{c}\% \text{ N derived from} \\ \text{storage-N + potential} \\ N_2\text{-fixation}\end{array} = 100 - \left[\frac{\text{atom } \%^{15}N \text{ excess, plant part}}{\text{atom } \%^{15}N \text{ excess, nutrient solution}} \times 100 \right]$$

The advantage of this method is that it is possible to calculate ($\%$N derived from storage-N plus potential N_2-fixation) for the major plant parts and combine this with the N-content data of Table 2 to provide an estimate of the N-balance of the plant system *without reference to the N-content of the nutrient solution*. This removes a potential major source of error, as the N in the nutrient solution has otherwise to be arbitrarily allocated between the surviving plants. Results of such calculations are given in Table 4. There is an indication of N_2-fixation in three out of the four variety/N-level treatments. The amounts estimated are about 17 to 24% of total-N in the plant, and are comparable to the amount obtained from the N-balance of zero-N treatments. Experiments with $^{15}N_2$ reported elsewhere in these volumes[6] provide support for the conclusion of N_2-fixation by sugar cane in solution culture.

The data for natural ^{15}N values of zero-N plants show that at the end of the experiment the roots of both varieties had the same atom $\% \, ^{15}N$ as did the seed piece stalks at the beginning, indicating no dilution by atmospheric nitrogen (0.3663 atom $\% \, ^{15}N$) such as would occur with N_2-fixation. Both tops and stalks of both varieties have reduced atom $\% \, ^{15}N$ showing the presence of some fixed nitrogen.

Calculations of fixation using natural ^{15}N values carry an inherently high error term, but even bearing this limitation in mind, it is nevertheless clear that the figures in Table 3 indicate a substantial proportion of fixed N_2 in the stalks and tops. Lack of evidence for fixed nitrogen in the roots should cause no surprise as the structual nitrogen of the roots will arise primarily from the stored nitrogen of the seed piece during the initial growth stages. Any N_2 subsequently fixed will remain in the seed piece or be preferentially translocated to the leaves. An analogous situation exists in soybean where the structural nitrogen of the root arises from soil nitrogen, subsequent nodule-derived nitrogen passing to the tops.[5]

Considering the two experiments, it is apparent that a certain small amount of N_2-fixation occurs without external inoculation when sugar cane is grown in solution culture, with zero or low NO_3-N. The amount is variable, but the data suggest about 12 to 20% of total plant nitrogen at the end of 60 days growth is due to fixation. It is notable that plants grown with zero-N can, through stored-N and fixed N_2, withstand the imbalance of the high levels of P, K, Ca, and Mg given in the culture solution. It is equally clear that much greater growth is achieved with an additional external nitrate source. Unlike nodulated legumes in solution culture, some additional nitrogen is necessary for really vigorous growth.

Table 3
ANALYSIS OF ATOM % ^{15}N IN PLANT PARTS AT HARVEST

Variety and Treatment						
	NA			**CB**		
Variety N-level	Zero N	3.5 ppm N	14 ppm N	Zero N	3.5 ppm N	14 ppm N
Tops, atom % ^{15}N	0.368[a]	0.759	2.150	0.369[a]	0.510	1.985
Roots, atom % ^{15}N	0.373[a]	0.891	1.458	0.374[a]	0.804	1.681
Stalk, atom % ^{15}N	0.369[a]	0.549	1.055	0.368[a]	0.481	0.890
Stalk, initial atom % ^{15}N		0.3725[a] ±0.001			0.374[a]	
Nutrient solution Atom % ^{15}N		4.040 ±0.002			4.040	

[a]   Natural ^{15}N values.

Table 4
DERIVATION OF NITROGEN IN PLANT PARTS CALCULATED FROM ISOTOPE DATA, AS PERCENT, AND AMOUNT PER PLANT (PLANTS RECEIVING COMBINED NITROGEN)

	3.5 ppm N		14 ppm N	
	NA	CB	NA	CB
Tops				
% N from storage + N₂-fixation	89.30	96.10	51.45	56.68
∴ amount of N storage + N₂-fixation, mg/plant	10.78	13.05	13.09	22.03
Roots				
% N from storage + N₂-fixation	85.72	88.09	70.29	64.22
∴ amount of N storage + N₂-fixation, mg/plant	9.93	8.08	20.30	15.15
Stalk				
% N from storage + N₂-fixation	95.00	96.88	81.26	85.75
∴ amount of N storage + N₂-fixation, mg/plant	14.72	12.96	14.23	20.95
Total N in plants, from storage + fixation, mg/plant	35.43	34.09	47.62	58.13
But, original N content of stalk, mg/plant	27.87	38.68	26.42	43.01
∴ amount due to fixation, mg/plant	+ 7.56	−5.41	+ 21.20	+ 15.12
and, expressed as % of total plant N	19.30	—	29.0	17.40

Although there have been some indications in other work that NA varieties supported better associative N₂-fixation than CB varieties, this is not really demonstrated in the present experiment. There is a possible indication that CB is not so efficient from the data showing loss of nitrogen from the seed piece (Table 5). It was already noted that, compared to NA, CB lost a greater proportion of its initial stalk weight at all N-levels, and similarly it shows a greater reduction in total N-content. This difference is not so marked when considered as percent of original N, although NA clearly used less stalk-N when N was increased in the nutrient solution. Nitrogenase activity was also reduced with increasing N-level, but it should be borne in mind that, e.g.,

Table 5

LOSS OF NITROGEN FROM "SEED PIECE" STALK, AND NITROGENASE ACTIVITY OF THE ROOTS AT HARVEST

N-level variety	Treatment					
	Zero N		3.5 ppm N		14 ppm N	
	NA	CB	NA	CB	NA	CB
Loss in N content of stalks, in 60 days, mg/plant	15.46	22.93	12.38	25.30	8.91	18.57
As % of original N content	53.2	57.4	44.4	65.4	33.7	43.2
Nitrogenase activity of roots at harvest, nmol C_2H_4/g/hr (mean of three determinations)	246	163	195	169	70	117

with NA, although ARA is greatly reduced with 14 ppm N the amount of roots is much greater, so the *total* nitrogenase activity per plant is not reduced so much.

Although NH_4^+ appears to be a good nitrogen source for the growth of sugar cane, the marked negative effect on acetylene reduction activity in the first experiment make it certain that even at low levels it is highly detrimental to nitrogen fixation. It takes little imagination to anticipate that high levels of ammonium fertilization in the field, whether given as ammonium sulphate or as aqua-ammonia, may also reduce severely, if not eliminate, potential N_2-fixation.

REFERENCES

1. Arias, O., Gatti, Irene, M., Silva, D. M., Ruschel, Alaides P., and Vose, P. B., Primeras observaciones al microscopio electrónico de bacterias fijadoras de N_2 en la raiz de la caña de azúcar, *Saccharum officinarum, Turrialba*, 28, 203, 1978.
2. Patriquin, D. G., Ruschel, Alaides, P., and Gracioli, L. A., Nitrogenase activity and bacterial infection of germinated sugar cane stem cuttings, in press.
3. Ruschel, Alaides P., Henis, Y., and Salati, E., Nitrogen-15 tracing of N-fixation with soil-grown sugar cane seedlings, *Soil Biol. Biochem.*, 7, 181, 1975.
4. Ruschel, Alaides P., Victoria, R. L., Salati, E., and Henis, Y., in Environmental role of nitrogen-fixing blue-green algae and asymbiotic bacteria, 1976, *Ecol. Bull. Stockholm*, 26, 297, 1978.
5. Ruschel, A. P., Vose, P. B., Victoria, R. L., and Salati, E., Comparison of isotope techniques and non-nodulating isolines to study the effect of ammonium fertilization on dinitrogen fixation in soybean, *Glycine max. Plant Soil*, 53, 513, 1979.
6. Ruschel, A. P., Matsui, E., and Vose, P. B., Potential N_2-fixation by sugar cane, *Saccharum* sp. seedlings in solution culture. II. Effect of inoculation; and dinitrogen fixation as directly measured by $^{15}N_2$, in *Associative N_2-Fixation*, Vose, Peter B. and Ruschel, A. P., Eds., CRC Press, Boca Raton, Fla., 1981.
7. Ruschel, A. P., unpublished.

Chapter 17

POTENTIAL N$_2$-FIXATION BY SUGAR CANE (*SACCHARUM* SP.) IN SOLUTION CULTURE.
II. EFFECT OF INOCULATION; AND DINITROGEN FIXATION AS DIRECTLY MEASURED BY ^{15}N$_2$

A. P. Ruschel, E. Matsui, E. Salati,* and P. B. Vose,**

TABLE OF CONTENTS

*   Centro de Energia Nuclear na Agricultura, Piracicaba, S. P., Brazil
**   International Atomic Energy Agency/United Nations Development Program.

I. INTRODUCTION

Nitrogenase activity has been found in sugar-cane stalks used for propagation, following germination,[6] while Patriquin et al.[3] found populations of microorganisms in stalks of sugar cane, related to nitrogenase activity, indicating that the presence of naturally occurring N_2-fixing bacteria is common. Further work using ^{15}N-labeled culture solution had indicated that these naturally occurring bacterial populations in the stalk were capable of fixing nitrogen without further inoculation, but part of the variability in that experiment could well have been due to different levels of the natural population.

Work with different varieties[6,7,8] had demonstrated differential effects on nitrogenase activity (NA) and $^{15}N_2$ incorporation[8] which suggests that modifications of the system might be brought about through inoculation. Döbereiner and Ruschel[1] observed N-enrichment in rice following inoculation with *Beijerinckia,* the same effect being obtained with two forage grasses *Paspalum* and *Cyperus* sp.,[4] while increased yield of sugar cane was obtained when *Azotobacter* culture was applied to the rootband of planting setts.[2]

In this paper a comparison is made of naturally occurring populations, inoculation with a mixture of N_2-fixing bacteria, and NO_3^--nitrogen, in two sugar cane varieties grown in solution culture.

Incorporation of nitrogen, and NA (acetylene reduction) were determined, together with two experiments involving direct measurements of dinitrogen fixation using $^{15}N_2$.

II. MATERIAL AND METHODS

Cuttings of two varieties of sugar cane, previously weighed, were placed in nutrient solution (KH_2PO_4—0.22 g, $MgSO_4 \cdot 7H_2O$ — 0.49 g, $CaCl_2$ — 0.17 g, micronutrients — 1.0 ml, FeEDTA — 10 ml, H_2O — 1000 ml) as previously described.[9] The treatments were: (1) control; (2) inoculation of cuttings with a mixture of microorganisms obtained by incubation of sugar-cane roots in N-free media; and (3) addition of 28 ppm of $N(KNO_3)$ to the nutrient solution.

Plants were harvested at 80 days and the weight determined, together with nitrogenase activity, of a portion of the roots separated from the stalks, as well as the original stalk. Nitrogen was determined by micro-Kjeldahl analysis.

The data were analyzed statistically by the Kruskal-Wallis nonparametric method of testing.

Two experiments were carried out to quantify N_2-fixation. In the first, one plant 60 days old was put in a chamber with $^{15}N_2$ added, as described in Ruschel et al.[8] and harvested 72 hr later, followed by analysis of ^{15}N enrichment of different parts of the plant (Table 1) by methods previously described.[5,8]

In the system for the second experiment, the roots were exposed to $^{15}N_2$ and the leaves to normal atmosphere. A diagram of this is shown in Figure 1. The plant was placed in a plastic vessel ($\simeq 3$ ℓ) containing nutrient solution and this vessel placed in a plastic bag. The upper part of the plastic bag was sealed with plastic foam to the plant and to the two plastic tubes used for air circulation, the open part of the bag being closed with a rubber band. The air from tank R was circulated by means of circulation pump P, entering through A and going out through B. To control CO_2 level, the gas was periodically circulated through the tube containing soda-lime, by means of valves S_1, S_2, and S_3. The gas pressure in the system was kept at the same pressure as the atmosphere, through the water level in Q.

By keeping the internal pressure identical to the external pressure it was hoped that

Table 1
NITROGEN FIXATION IN SUGAR CANE GROWN IN
NUTRIENT SOLUTION FREE OF MINERAL-N AFTER 72 HR IN
$^{15}N_2$ ATMOSPHERE (EXPERIMENT 1)

Treatment	Part of plant	Weight (g)	Total N (mg)	$^{15}N\%$ ± 0.003	^{15}N excess	μg N-fixed
Uninoculated	Leaves	2.495	13.97	0.367		
Control	Roots	1.523	9.75	0.367		
Plant	Stalks	4.384	32.00	0.371		
Uninoculated	Leaves	4.127	30.13	0.382	0.015	12
$^{15}N_2$	Roots	2.088	17.12	0.635	0.268	42
Plant	Stalks	6.156	52.33	0.593	0.222	109

Note: p $^{15}N_2$ — 35.3%; O_2 = 12.86; Ar = 21.04 and CO_2 0.63.

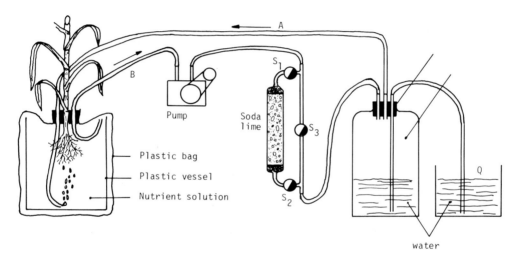

FIGURE 1. Diagram of the experiment with sugar-cane roots exposed to $^{15}N_2$ with leaves in the normal atmosphere.

no great exchange with the outside air would occur, as observed in preliminary experiments using argon. However, there was a relatively large leakage of the system during the experiment, which caused dilution of ^{15}N. Further tests indicated that there was considerable diffusion of the gas through the plastic bags used in the experiment. Besides, young sugar-cane plants have porosity in the longitudinal direction, which contributes to gas exchange with the outside.

III. RESULTS AND DISCUSSION

Tables 2 to 4 show results of weight, nitrogenase activity, total N, and increase in nitrogen. Tables 1 and 5 show N_2-fixed as measured by ^{15}N-enrichment, from the experiments with plants exposed to 15-dinitrogen.

With addition of NO_3^--N there was increased weight in both varieties as well as with inoculation, in the latter case not quite statistically significant. As in the previous experiments,[9] plants grown with NO_3^- greatly exceeded the weight of other plants, even when inoculated. On observing the N-content of the plants (Table 3) it can be seen

Table 2

EFFECT OF INOCULATION OF A MIXTURE OF N₂-
FIXING BACTERIA AND NO₃-N ON THE WEIGHT (G)
OF PLANT PARTS OF TWO VARIETIES OF SUGAR
CANE GROWN IN NUTRIENT SOLUTION (AVERAGES
OF 6 REPLICATES)

Variety	Treatment	Leaves	Roots	Stalks	Σ
	Without inoc.	3.514 b[a]	2.240 b	4.512	10.266 b
CB41-76	Inoculated	3.682 b	2.660 b	5.240	11.582 b
	+ N(NO₃)	18.220 a	9.588 a	5.295	33.103 a
	Without inoc.	1.800 b	1.920 b	5.533	9.253 b
NA56-79	Inoculated	2.655 ab	2.373 ab	6.364	11.392 ab
	+ N(NO₃)	11.397 a	5.033 a	5.253	21.683 a

[a]   In the same column and variety, values followed by different letters are statistically different.

Table 3

TOTAL N AND INCREASE OF N (MG) DUE TO N₂-FIXATION IN SUGAR
CANE, INOCULATED AND WITH N(NO₃), GROWN IN NUTRIENT
SOLUTION FREE OF MINERAL N (MEAN OF 6 REPLICATIONS)

		Nitrogen (mg)					N-increase	
Variety	Treatment	Roots	Leaves	Stalks	Σ	Initial N in stalks	Whole plant	Leaves and stalks
	Without inoc.	13.14 b[a]	24.50 b	19.19	56.83 b	29.12	28.31 b	8.80 b
CB41-76	Inoculated	10.76 b	26.56 b	23.04	60.36 b	29.52	30.84 b	7.85 b
	+ N(NO₃)	40.88 a	122.41 a	24.77	188.06 a	30.32	157.74 a	132.97 a
	Without inoc.	12.13 b	14.43 b	22.59	49.15 b	37.30	16.20 b	1.91 c
NA56-79	Inoculated	39.69 a	17.59 b	30.57	87.85 ab	39.39	48.45 ab	18.67 b
	+ N(NO₃)	44.85 a	64.27 a	22.08	131.20 a	35.66	93.55 a	71.46 a

[a]   In the same column and variety, values followed by different letters are statistically different.

Table 4

EFFECT OF INOCULATION WITH A MIXTURE OR N₂-
FIXING ORGANISMS AND NO₃-N ON NITROGENASE
ACTIVITY (ARA) OF ROOTS AND ORIGINAL STALKS OF
TWO VARIETIES OF SUGAR CANE (CB41-76 AND NA56-79).
MEAN OF 6 REPLICATES

		nmol C₂H₄/g/root/ hr		nmoles C₂H₄/plant/ hr	
Variety	Treatment	Root	Stalks	Root	Stalks
	Without inoc.	73.7 b[a]	43.3 b	168.5 b	188.7 b
CB41-76	Inoculated	95.1 ab	75.5 ab	257.5 ab	379.3 ab
	+ N(NO₃)	151.0 a	140.4 a	1451.8 a	716.5 a
	Without inoc.	54.7 b	82.3 b	101.0 b	431.7 b
NA56-79	Inoculated	142.4 a	226.0 a	329.8 a	1519.0 a
	+ N(NO₃)	88.6 ab	54.7 ab	479.0 a	269.9 b

[a]   In the same column and variety, values followed by different letters are statistically different.

Table 5
NITROGEN FIXATION AND NITROGENASE ACTIVITY IN SUGAR CANE (80 DAYS) GROWN IN NUTRIENT SOLUTION WITH AND WITHOUT N(NO₃), AND WITH AND WITHOUT INOCULATION (SECOND EXPERIMENT)

Treatment	Part of plant	Weight (g)	Total-N (mg)	nmol C_2H_4/ hr	$^{15}N\%$ + 0.003	^{15}N excess	N_2-fixed (μg)
Control	Leaves	5.234	30.88	24	0.365		
	Roots	4.752	41.34	178	0.365		
	Stalks	5.234	23.55	293	0.367		
$^{15}N_2$ non-inoc. plant	Leaves	2.600	21.06	0	0.367	0.002	6.6
	Roots	2.503	25.03	0	0.372	0.007	27.4
	Stalks	6.79	26.48	217	0.394	0.027	111.7
$^{15}N_2$ inoc. plant	Leaves	2.826	22.89	0	0.365	0	$\sim$0
	Root (plant)	0.839	8.81	116	0.379	0.014	19.3
	Stalks (plant)	1.969	7.68	797	0.398	0.031	37.2
	Root[a]	1.686	11.46	23	0.380	0.015	26.9
	Stalks[a]	4.075	15.08	1399	0.409	0.042	103.7
$^{15}N_2$ N(NO₃) plant	Leaves	8.728	70.70	0	0.366	0.001	$\sim$0
	Roots	8.084	76.80	116	0.387	0.022	246.0
	Stalks	5.430	21.18	1295	0.468	0.101	334.2

[a] Parts separated from plant before starting the experiment.

that inoculation triples the total N of roots compared with noninoculated plants. Leaf nitrogen was affected by NO_3 addition only. The initial N content of the stalks was calculated (Table 3), taking into account that the conversion factor from fresh weight to dry weight of the original stalks was 4.63 and 4.32, and that the initial percent N of NA and CB was 0.65 and 0.57%, respectively. It was then possible to determine the increase in N due to biological fixation, and the N-uptake from the NO_3 treatment, with or without the original stalk (Table 3 — increase in N).

Cumulative N indicates that N_2-fixation was common for both varieties, the N-uptake of uninoculated and inoculated plants being higher in CB41-76 than NA56-79. Nitrogenase activity was found in all plants (Table 4) including treatments with NO_3, always being higher in inoculated and NO_3 treatments, except for the stalks of NA56-79. This indicates that N_2-fixation occurred even when NO_3 was included in the nutrient solution. Indeed, with CB41-76 Table 4 shows a clear stimulating effect of NO_3-nutrition on nitrogenase activity compared with either inoculated or uninoculated zero-NO_3 plants. This contrasts with the previous experiment,[9] when nitrogenase activity was apparently reduced with increased NO_3, although the difference was not great, especially with CB41-76.

Direct $^{15}N_2$-fixation was evaluated in plants from all treatments (Tables 1 and 5). ^{15}N appeared in leaves only in the experiment in which plants were exposed in a chamber with $^{15}N_2$ for 3 days. When plants had only the roots exposed to $^{15}N_2$ for only 2 days, there was no ^{15}N incorporation in the leaves. This is probably just a time factor, but the possibility that microorganisms present in leaves or near the original stalks could fix nitrogen, cannot be excluded. The high N-content of NA56-79 roots indicated that probably the fixed N_2 was in a sink, retained presumably by the bacteria to be distributed later. A time course experiment is needed to determine this.

The fixation of N_2 by sugar cane in culture solution was confirmed. The effect of inoculation is shown, clearly, in terms of increased N-content and greater nitrogenase (NA) activity in NA56-79, but not in CB41-76. No effect of inoculation was found in the short term $^{15}N_2$ experiment. The differing effect of inoculation between varieties

I. INTRODUCTION

Nitrogen fixation has been detected in *Gramineae*, and differences among cultivars and ecotypes in relation to the occurrence of the N$_2$-fixing bacteria or nitrogenase activity have been shown.[4] However, it is believed that the N$_2$-fixing capacity is due to an interaction between genotype and the environment. In order to analyze the genotypic effect it must be isolated.

Sugar cane is propagated vegetatively. Every bud of a clone or variety has the same genetic constitution, which makes it easier to separate a genetic effect from the phenotype without handling pure lines. Sugar cane is highly heterozygous and by crossing two varieties the genetic recombination gives rise to large variability among the F$_1$ produced, which enhances the possibilities of selection for particular characteristics.

Nitrogenase activity in the sugar-cane rhizosphere has been established[3,7] with N$_2$-fixation being proved by ^{15}N$_2$ incorporation in the plant.[8,9]

In the present paper 2-months old sugar-cane plants, originating from cuttings (vegetative propagation) and seedlings germinated from true seeds were analyzed for nitrogenase activity by the acetylene reduction method (ARA), and N$_2$-fixing ability under ^{15}N$_2$ atmosphere, to detect and to isolate a potential genotypic effect on nitrogen fixation in cane.

II. VARIETAL DIFFERENCES AFFECTING NITROGENASE ACTIVITY IN THE RHIZOSPHERE

Direct evidence of a varietal effect on nitrogen fixation by *Gramineae* was reported by Ruschel et al.,[9] from analysis of the ^{15}N$_2$ fixed in the roots, tops and cuttings of three sugar-cane varieties exposed to ^{15}N$_2$ atmosphere in a chamber (Table 1). The fixed nitrogen in the roots was translocated to the leaves. The results show that variety CB46-47 on average fixes more nitrogen during 24-hr incubation than CB41-76 and CB47-355. All these three varieties are progenies of POJ2878 as female parent crossed with an unknown pollen donor parent.

Using the acetylene reduction method,[5] Ruschel and Ruschel[10] measured the nitrogenase activity of different parts of cane: whole plants plus soil in an intact system; whole plants with roots free of soil; roots plus shoots without cuttings; cuttings used for germination (roots and shoots removed); external and internal parts of cuttings; and detached roots. Except for the first of these, all others were considered to be disturbed systems. Ethylene was analyzed 1, 2, and 4 hr after acetylene injection. Six of the most commonly planted sugar cane varieties in the state of São Paulo were used in this study: CB41-76, CB46-47, CB47-355, CP51-22, IAC51/205 and NA56-79, the data being averages of five plant replicates.

The results presented in Table 2 show evidence of a varietal effect on N$_2$-fixing (ethylene) activity of microorganisms in sugar cane. Varieties NA56-79 and CB46-47 had, on average, the greatest nitrogenase activity, which in the intact system (normal oxygen atmosphere), differed statistically from the others. In a low oxygen atmosphere CB47-355 as well as CB46-47 and NA56-79 showed high nitrogenase activity. Variety CB41-76 had the lowest nitrogenase activity in all experiments.

It is already known that the nitrogen concentration in sugar-cane leaves varies according to the genotype.[6] These authors carried out a nutritional survey by foliar analysis and found a varietal effect on leaf nitrogen content, independent of soil type. A high correlation was obtained between concentrations of nitrogen found by these authors in five of the six varieties studied on the same soil, and the average volume of ethylene evolved from these varieties in the intact system under normal oxygen from

Table 1
ENRICHMENT OF NITROGEN (μg ^{15}N FIXED IN 24 HR) IN ROOTS, TOPS, AND CUTTINGS OF THREE VARIETIES OF SUGAR CANE. AVERAGES OF THREE REPLICATES

Varieties	Roots	Tops	Cuttings	Averages
CB46-47	0.161	0.085	5.142	1.796 a[a]
CB41-76	0.126	0.358	2.788	0.091 b
CB47-355	0.208	0.488	1.367	0.674 b
Averages	0.165 b	0.297 b	3.099 a	

[a] Values followed by the same letter do not differ statistically by 5% probability.

Table 2
NITROGENASE ACTIVITY (NMOL ETHYLENE EVOLVED/PLANT 2-MONTHS OLD) OF DIFFERENT PARTS OF SUGAR CANE. AVERAGED MEASUREMENTS, TAKEN AFTER 1, 2, and 4 hr OF EXPOSURE TO ACETYLENE (AVERAGES OF FIVE PLANTS)

Varieties	A	B	C	D	E
CB41-76	84.9 b*[a]	66.4 b	32.1 b	—	—
CB46-47	551.1 a	1485.8 a	375.6 a	248.2	313.4
CB47-355	159.1 b	1033.2 a	29.5 b	76.3	417.5
CP51-22	214.7 b	482.2 b	226.3 ab	—	—
IAC51/205	299.4 b	449.4 b	75.4 b	—	—
NA56-79	683.6 a	1754.9 a	415.9 a	152.5	613.1
Averages	332.1	878.5	192.5	159.0 b	466.0 a

[a] = In the same column, values followed by different letters are statistically different at 5% level of significance.

Note: A = Whole plant plus soil (intact system) — air O_2 atmosphere.
B = Whole plant plus soil in low O_2 atmosphere.
C = Whole plant with roots free of soil (disturbed system).
D = Roots plus top part without cuttings (disturbed system).
E = Cuttings used for germination (roots and shoots removed).

air ($r = 0.821$, on the verge of significance at the 5% level). This positive correlation might be an indication that the differential ability of 2-months old plants to fix nitrogen might influence the N-balance of these varieties.

It is interesting to note that mature stalks of varieties CB46-47 and NA56-79 have high sucrose content, while CB41-76 has a low sucrose content, although it has good productivity.[1]

Nitrogenase activity measured in intact systems seems to be greater than in disturbed (Table 2) as shown by Ruschel.[7] This indicates that sugar cane influences the soil activity of N_2-fixing populations, which is in accord with Döbereiner[2] who observed an increase of N_2-fixing bacteria in the cane soil rhizosphere.

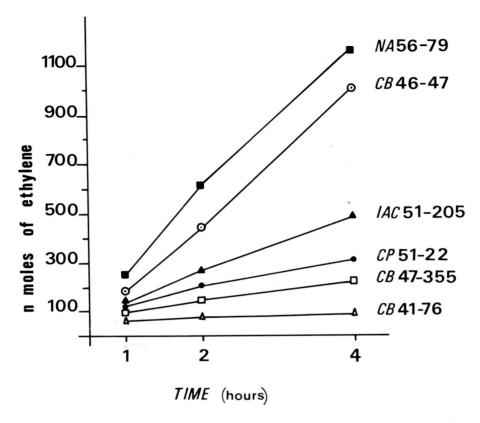

FIGURE 1.   nmol of ethylene evolved per plant, 2-months of age in intact system, after 1, 2, and 4 hr of exposure to acetylene.

The volume of ethylene produced after 4 hr of exposure to acetylene (Figure 1) was higher than that measured at earlier intervals, indicating that the rates of fixation in varieties NA56-79 and CB46-47 were similar during the experimental period, but decreasing in the other varieties. This probably indicates a more efficient N_2-fixing system in the two cited varieties.

Ruschel et al.,[8] working with 3-months old cane seedlings under $^{15}N_2$ atmosphere, noticed that the sites of N_2-fixing were in the roots. However, results shown in Tables 2 and 3 indicate that germinated cuttings of some varieties from which the roots and shoots were excised, had a higher N_2-fixation and nitrogenase activity than the other parts of the plant. There appear to be differences, not statistically significant, in nitrogenase activity among the internal part of the cuttings, the external part (rind) and the roots (Table 3), the rind showing the highest activity.

Investigations need to be made on the carbohydrate content of germinated cuttings, the nature of exudates around them, and their effects on N_2-fixing microorganisms. Varieties, such as CB46-47 and NA56-79, that maintain a high level of nitrogenase, even without the "seed" stalk, are desirable since it is not yet known how the stalks are able to cause proliferation of the N_2-fixers.

III. EFFECT OF VARIETY ON NITROGEN FIXATION IN SEEDLING PROGENIES

Ruschel et al.[12] measured nitrogenase activity (ARA), in 40 weight-groups of five seedlings germinated from seeds of CB41-76 and Co617 sugar cane varieties in an intact

Table 3
NITROGENASE ACTIVITY
(NMOL ETHYLENE EVOLVED/
PARTS OF A SINGLE PLANT 3-
MONTHS OLD) OF
GERMINATED CUTTINGS AND
ROOTS OF SUGAR CANE
AFTER FOUR HOURS OF
EXPOSITION TO ACETYLENE
(AVERAGES OF THREE
PLANTS)

Varieties	A	B	C
CB46-47	834.9	22.0	7.9
CB47-355	192.1	188.4	22.0
IAC51/205	427.9	619.0	36.7
NA56-79	314.4	787.8	73.4
Averages	242.3	404.3	35.0

Note: A = Internal part of germinated single node cuttings.
B = External part (rind) of germinated single node cuttings.
C = Detached roots.

system of whole plants plus soil. The results (Table 4) showed that nitrogen fixing ability was greater in the seedlings germinated from seeds harvested from the CB41-76 variety than in the seedlings originating from Co617.

The inflorescences of these varieties were exposed and received pollen from many other varieties in a polycross area. As the seeds originated from a polycross a selfing or a maternal effect can be postulated, in order to explain the conspicuous difference observed between the two groups of seedlings. The high range of variability presented makes it possible to select for high nitrogen fixation ability, although it would be necessary to determine that the response of seedlings for this ability is correlated with that of the adult plant.

IV. INHERITANCE OF NITROGEN FIXATION POTENTIAL

Two crosses (CP36-105 × CP38-34 and Co331 × Co290) have been studied in the F_1 generation (Ruschel et al.[13]) and the results are reported here. The two crosses differ greatly with respect to the origin of the parent. The background of Co331 and Co290 includes not only the same varieties but also the same crosses. CP36-105 and CP38-34, on the other hand, have largely different backgrounds, CP38-34 includes several *Saccharum spontaneum* as well as one cross with the Panshai group of cane.

Germinated cuttings from F_1 and F_2 and clones from F_1 seedlings were analyzed for nitrogenase activity in an intact system (plant plus soil medium) in a sealed jar with 0.1 atm of C_2H_2. One series (A) was planted in soil and analyzed 2, 3, and 4 months after rooting; the data shown in Figure 2 being the average of these three measurements. The second series (B) represents cuttings of the same material 1 year later, planted in sterile vermiculite plus sand and analyzed 2, 3, and 4 weeks after rooting, the data in Figure 2 again being averages. The F^a_1 and F^b_1 of cross CP36-105 × CP38-34 represent clones originating from different F_1 seeds.

Table 4

NMOL OF ETHYLENE EVOLVED PER GROUP OF FIVE SEEDLINGS OF SUGAR CANE, AFTER 1, 2, 3, AND 4 HR OF EXPOSURE TO AN ATMOSPHERE OF 10% ACETHYLENE. MEASUREMENTS ON THREE DIFFERENT DAYS

Date of measurement	Varieties	Hr	nmol[a]		
		1	1,402.3	±	281.0
Oct. 6	CB41-76	2	4,414.5	±	850.5
		3	10,528.5	±	1,475.4
		1	211.0	±	103.0
Oct. 6	Co617	2	302.1	±	58.9
		3	703.7	±	222.1
		1	207.8	±	122.8
Oct. 9	CB41-76	2	688.0	±	377.3
		3	2,062.3	±	1,196.1
		1	78.6	±	37.6
Oct. 9	Co617	2	85.6	±	35.8
		3	193.8	±	105.0
		1	126.0	±	59.2
Oct. 24	CB41-76	3	858.4	±	181.9
		4	1,261.6	±	562.8
		1	25.2	±	16.5
Oct. 24	Co617	3	64.6	±	36.2
		4	67.7	±	34.3

[a]   Oct. 6 measurement — averages of 24 groups of 5 seedlings.
Oct. 9 measurement — averages of 8 groups of 5 seedlings.
Oct. 24 measurement — averages of 8 groups of 5 seedlings.

It will be observed that the results are consistent at the two sampling dates. In the case of the cross Co331 × Co290 the F_1 plants show ethylene production similar or a little greater than the parents. Cross CP36-105 × CP38-34 involved parents with widely differing capacity for nitrogenase activity. The F_1 progeny of this cross show much lower nitrogenase activity than the parent with the highest activity.

These results are considered to be the first tangible and promising evidence of N_2-fixing potential in sugar cane being inherited in an ordinary fashion accessible to ordinary breeding methods. Though it is too early to speculate on the details, cross CP36-105 × CP38-34 seems to demonstrate partial dominance of low N_2-fixing ability.

V. CONCLUSIONS

From the data presented, it can be concluded that there are differences in the nitrogen fixation ability of sugar-cane varieties. These differences, with wide ranges of variation, are also found in progenies, potentially permitting the screening for this capacity with maximum effect.

It was also evident that the inheritance mechanism of the factors that control nitrogen fixation in sugar cane occurs in a normal fashion, which suggests that ordinary breeding methods can be applied to enhance this character.

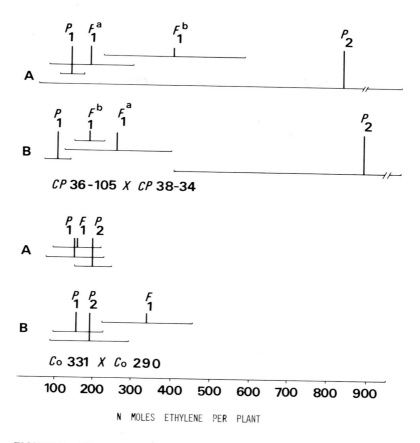

FIGURE 2. Nitrogenase activity (nmol of ethylene per plant) of crosses CP36-105 × CP38-36 and Co331 × Co290 in two experiments, A and B.

REFERENCES

1. **Bassinello, A. I.,** Apreciacão sobre experimentos de competição de variedades da šerie 71, *Bras. Acucareiro,* 87(5), 42, 1976.
2. **Döbereiner, J.,** Influência da cana-de-açúcar na população de *Beijerinckia* do solo, *Rev. Bras. Biol.,* 19, 251, 1959.
3. **Döbereiner, J., Day, J. M., and Dart, J.,** Nitrogenase activity in the rhizosphere of sugar cane and other tropical grasses, *Plant Soil,* 37, 191, 1972.
4. **Döbereiner, J.,** Plant genotype effect on nitrogen fixation in grasses, in *Genetic Diversity in Plants,* Ed. B. Hollander, Plenum Press, New York, 1977, 325.
5. **Hardy, R. W. F., Holstein, R. D., Jackson, E. K., and Bruns, R. C.,** The acetylene ethylene assays for N₂ fixation: laboratory and field evaluation, *Plant Physiol.,* 43, 1185, 1968.
6. **Orlando Fo., J. and Haag, H. P.,** Influência varietal e do solo no estado nutricional na cana-de-açúcar (*Saccharum* spp.) pela análise foliar, *Bol. Téc. no. 2 da Coordenadoria Regional Sul.,* PLAN-ALSUCAR, Araras, 1976, 52.
7. **Ruschel, A. P.,** Fixação Biologica de Nitrogênio em Cana-de-Açúcar, Ph.D. thesis, ESALQ-University of Paulo, Brasil, 1975.

8. **Ruschel, A. P., Henis, Y., and Salati, E.,** Nitrogen-15 tracing of N-fixation with soil grown sugar cane seedlings, *Soil Biol. Biochem.,* 7, 181, 1975.

9. **Ruschel, A. P., Victoria, R. B., Salati, E., and Henis, Y.,** Biological Nitrogen Fixation in Sugar Cane (*Saccharum officinarum*L.), in Ecological Bulletins/NFR 26: Environmental Role N-Fixing Blue-green Algae and Assymbiotic Bacteria Granhall, U., Ed., Swedish University of Agriculture, Forestry and Veterinary Medicine, Uppsala, Sweden, 1978.

10. **Ruschel, A. P. and Ruschel, R.,** Varietal differences affecting nitrogenase activity in the rhizosphere of sugar cane, *Proc. Soc. Sugar Cane Technol.,* Reis, F. S., and Dick, Eds., 1977

11. **Ruschel, A. P.,** Fixação biológica do nitrogênio, in *Fisiologia Vegetal I,* Ed. Universidade de São Paulo, Brasil, 1978, 167.

12. **Ruschel, R., Cesnik, R., and Ruschel, A. P.,** Fixação de Nitrogênio em Plântulas de Cana-de-Açúcar, 10th Congr da Soc. dos Técnicos Açucareiros do Brasil, Maceió, Al., Brasil, 1979.

13. **Ruschel, R., Ruschel, A. P., and Blixt, S.,** Genetics of N₂-fixing potential in sugar cane (*Saccharum* sp.), 1979, unpublished.

Chapter 19

RADIORESPIROMETRY STUDIES AS AN INDICATION OF SOIL MICROBIAL ACTIVITY IN RELATION TO THE ROOT SYSTEM IN SUGAR CANE, AND COMPARISON WITH OTHER SPECIES

J. R. Freitas, A. P. Ruschel,* and P. B. Vose**

TABLE OF CONTENTS

*   Centro de Energia Nuclear na Agricultura, Piracicaba, S. P., Brazil.
** International Atomic Energy Agency/United Nations Development Program.

I. INTRODUCTION

Sugar cane affects the population of N_2-fixing microorganisms in the rhizospehre,[4] Hartt et al.[6] found that labeled sucrose appears to move out of cane roots, suggesting carbohydrate exudation, while Ruschel et al.[8] demonstrated enhancement of N_2-fixation in soil and roots following glucose addition to soil. Döbereiner[3] had observed that cane favors N_2-fixing bacteria, especially *Beijerinckia,* due to incorporation of sucrose from cane in the soil, and working with corn Döbereiner and Lemos[2] demonstrated that addition of sucrose to the soil decreased the weight of plants more when mineral-N was not added. However, sucrose increased the N_2-fixing ability of *Beijerinckia,* being higher with *B. indica* than with *B. fluminensis.*

The evaluation of heterotrophic populations can be achieved indirectly by radiorespirometry, using glucose-^{14}C to assess respiration activity of soil samples.[9] In this paper the effect of sugar cane on the activity of soil microflora was studied in comparison with other plants.

II. MATERIAL AND METHODS

Samples of Red Latosol soil (Terra Roxa Estruturada-TRE) from soil planted with different crop species (14-month old ratoon sugar cane, native and 10-year old planted *Pinus caribaea* forest, 60-day old maize, *Zea mays,* and 40-day old beans, *Phaseolus vulgaris*) were taken directly from the field. Superficial soil (1 cm layer) was removed and 10 subsamples of 10 g each were mixed to obtain the sample which was placed in a glass flask and immediately analyzed in the laboratory. Samples from the rhizosphere had the roots carefully taken out. In the last experiment the soil samples from ratoon cane were taken at different depths and distances with four replicates, but with eight replicates when the distance was 35 cm.

Two grams of soil were put in the respirometry flasks (an Erlenmeyer with a small tube supported by a stainless steel wire in the stopper), then moistened, and 0.5 mℓ of glucose (labeled with glucose-^{14}C) of different concentrations was added, followed by incubation for 1 hr, when the reaction was stopped by 2 mℓ N HCl. $^{14}Co_2$ evolved was absorbed by phenylethylamine contained in the small tube, which was later transferred quantitatively to scintillation solution. Pulses of ^{14}C from samples, and the original ^{14}C-glucose added, were determined in a Beckman ® LS-230 liquid scintillation counter. The development of the procedure is described in detail elsewhere.[10]

III. RESULTS AND DISCUSSION

Results in Tables 1 and 2 present the amount of glucose consumed by respiratory activity of microorganisms in the soil samples, and show the effect of different crop species (sugar cane, native and pinus forest, corn, and *Phaseolus* beans) on the intensity of microbial activity.

Comparison of the soil between rows and the rhizosphere soil was made, bearing in mind that rhizosphere is the region of the soil in which plants exert their major influence. From Table 2 it can be seen that the rate of microbial respiration in the sugarcane rhizosphere is affected by the addition of glucose to the same degree as the forestry rhizosphere, when compared to soil from between the rows of cane. Kibani et al.[7] studying the rhizosphere effect of maize and soybean, intercropped or in monoculture, noticed that intercropping increased the bacterial rhizosphere to soil (R/S) ratio of maize and soybean, whereas fungal R/S decreased with intercropping in comparison to monoculture conditions, the rhizosphere always having an increased micro-

Table 1
EFFECT OF LEVELS OF ADDED GLUCOSE ON MICROBIAL ACTIVITY IN RELATION TO PLANT COVER AND RHIZOSPHERE

Added glucose (μmol/ml)		Glucose consumed (nmol/g/hr)				Average (treatment)
		1	2	5	10	
Expt. 1 (before rain)	Natural forest	37 e	49 cd	46 d	56 ab	47.2 a
	Pinus forest	39 c	45 d	48 cd	52 bc	46.0 a
	Sugar cane (rhizosphere)	25 f	33 e	49 cd	60 a	41.6 a
	Sugar cane (between rows)	12 h	8 i	14 g	25 bc	14.9 b
	Average (level)	28.2c	33.9bc	39.4cd	48.3a	
Expt. 2 (after rain)	Sugar cane (rhizosphere)	43 c	30 cd	154 a	89 b	79.0 a
	Sugar cane (between rows)	13 cd	19 cd	12 d	23 cd	16.8 b
	Average (level)	28.0cd	38.5c	83.0b	97.5a	

Note: Means with different letters in the same line or column of averages are significantly different at 5% level of significance.

Table 2
PLANT SPECIES AND RHIZOSPHERE EFFECT ON HETEROTROPHIC ACTIVITY OF SOIL

Treatments	Glucose consumed (nmol/g/hr)			
	Sugar cane	Maize	*Phaseolus* (beans)	Average (treat)
Rhizosphere	67	71	31	56.3 a
Between rows	35	53	29	39.0 ab
Average (crop)	51.0 ab	62.0 a	30.0 b	

Note: Means with different letters in the same line or column of averages are significantly different at 5% level of significance.

bial population. In the present case, the forestry rhizosphere can be presumed to be in equilibrium in relation to the activity of microorganisms, since the turnover of organic material is continuous and the disturbance is small. As sugar cane has microorganisms similar to the forest rhizosphere this suggests that root exudation of organic substances is high.

This fact is emphasized by the results of Experiment 2 (Table 1) where the activity of sugar-cane soil in the rhizosphere increased after rain, comparing the results with Experiment 1, carried out during a dry period. Döbereiner and Alvahydo[5] noted the effect of leachates from sugar-cane leaves on N$_2$-fixing microorganisms, but Brown[1] studying the rhizosphere of *Paspalum notatum* and *Azotobacter paspali* found that the effect of the bacteria was to produce growth factors that influenced plant growth, more than through a nitrogen fixation effect. Again, Table 2 shows the greater effect of sugar cane and maize rhizosphere on the activity of heterotrophs, compared with the *Phaseolus* rhizosphere, demonstrating the strong influence of these C$_4$ plants.

The distribution of microbial activity in the soil of ratoon cane can be seen in Figure 1, the activity being highest 45 cm from the stalks, where the roots are very active in nutrient absorption, the microbial activity decreasing with depth.

It is clear that sugar cane, besides having an active system of N$_2$-fixation, also influences the whole natural population of soil microorganisms.

DEPTH (cm)

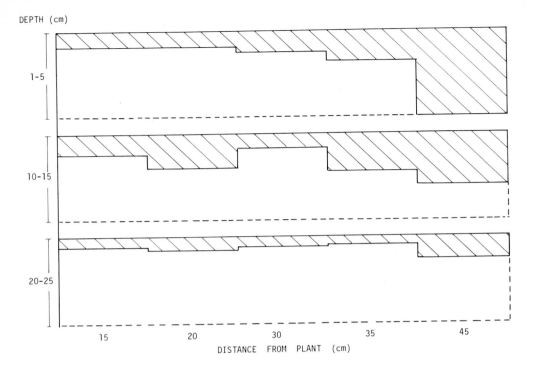

FIGURE 1.   Relative microbial activity in soil samples of ratoon cane at different depths and distances.

REFERENCES

1. **Brown, M. E.,** Role of *Azotobacter paspali* in association with *Paspali notatum, J. Appl. Bacteriol.,* 40, 341, 1976.
2. **Döbereiner, J. and Lemos, P.,** Acão da sacarose e de *Beijerinckia* na fixação de Nitrogênio e agregação do solo, *Bol. do Inst. de Ecologia e Experimentação Agricolas,* Brasil, No. 20, 1958.
3. **Döbereiner, J.,** A ocorrência de *Beijerinckia* em alguns Estados do Brasil, *Rev., Bras. Biol.,* 19(2), 151, 1959.
4. **Döbereiner, J.,** Influência da cana-de-açucar na população de *Beijerinckia* do solo, *Rev. Bras. Biol.,* 19, 251, 1959.
5. **Döbereiner, J. and Alvahydo, R.,** Sobre a influência da cana-de-açucar na ocorrência de *Beijerinckia* no solo. II. Influencia das diversas partes do vegetal, *Rev. Bras. Biol.,* 19, 401, 1959.
6. **Hartt, C. E., Kortschak, H. P., Forbes, A. J., and Burr, G. O.,** Translocation of ¹⁴C in sugar cane, *Plant Physiol.,* 39, 305, 1963.
7. **Kibani, T. H. M., Keswani, C. L., and Chowdhury, M. S.,** Rhizosphere populations in intercropped maize and soybeans, in Report of a Symp. IDRC, Monyo, J. H., Ker, A. D. R., and Campbell, M., Eds., 1976, 13.
8. **Ruschel, A. P., Victoria, R. L., Salati, E., and Henis, Y.,** Nitrogen fixation in sugar cane (*Saccharum officinarum* L.), environmental role of nitrogen-fixing blue-green algae and assymbiotic bacteria, *Ecol. Bull. Stockholm,* 26, 297, 1978.
9. **Mayaudon, J.,** Use of radiorespirometry in soil microbiology and biochemistry, in *Soil Biochemistry,* Vol. 2, Marcel Dekker, New York, 1971, 202.
10. **Freitas, J. R., Vose, P. B., Nascimento Fo., V. F., and Ruschel, A. P.,** Estimativa da atividade da microflora heterófica em solo terra roxa estruturada usando respirometria com glicose -¹⁴C, *Energ. Nucl. Agric. Piracicaba,* 1(2), 123, 1979.

Chapter 20

DETERMINATION OF POTENTIAL N$_2$-FIXING ACTIVITY OF BACTERIA IN SUGAR-CANE ROOTS AND BEAN NODULES USING TRITIATED ACETYLENE REDUCTION TECHNIQUE AND ELECTRON MICROAUTORADIOGRAPHY

D. M. Silva, A. P. Ruschel, E. Matsui, N. de L. Nogueira,* and P. B. Vose**

TABLE OF CONTENTS

*   Centro de Energia Nuclear na Agricultura, Piracicaba, S. P., Brazil.
** International Atomic Energy Agency/United Nations Development Program.

I. INTRODUCTION

Biological nitrogen fixation in the rhizosphere of *Gramineae* has been demonstrated by evaluation of nitrogenase activity, e.g.,[3] and 15-dinitrogen studies with sugar cane, *Saccharum* sp.[5,6] and tropical grasses.[2] However, it has not been clear whether such N₂-fixation can take place specifically within the root, as opposed to the rhizoplane or the rhizosphere,[7] even through the presence of bacteria may have been demonstrated in root cells by means of light or electron microscopy. Presence is no proof of N₂-fixation.

Previous methods of determining N₂-fixation by bacteria in the root have been indirect, based on determining acetylene reduction activity (ARA) on excised incubated roots which had been previously surface sterilized with mercuric chloride or chloramine-T. This paper describes an attempt to develop a direct method for confirming N₂-fixation within the root. The problem is common to all associative N₂-fixation systems but we have been particularly concerned with sugar cane.

II. EXPERIMENTAL

The acetylene reduction technique usually used to determine nitrogenase activity suggested the use of tritium labeled acetylene, combined with the use of high resolution microautoradiography to record the activity of potentially N₂-fixing bacteria. There was some doubt as to whether tritium labeled ethylene, the expected product of tritiated acetylene reduction, could be "fixed", but a preliminary experiment has been carried out using sugar-cane roots and *Phaseolus vulgaris* bean nodules, and the first results seem to be promising.

The following samples were placed for 30 min in an atmosphere containing 10% tritiated acetylene: (1) approximately 0.5 × 1 cm long segments of field-grown sugar-cane roots cut 1.5 cm from root cap, (2) similar segments incubated for 24 hr in a N-free culture medium to facilitate the development of N₂-fixing bacteria, and (3) *Phaseolus* bean nodules as a control. Due to the universal presence of the bacteria it was found impossible to obtain control sugar-cane roots free of bacteria, except through a very involved tissue culture technique and which has so far not been developed sufficiently to produce mature sterilized roots of corresponding age.

The tritiated acetylene was obtained by reacting CaC_2 with ^3H_2O in previously evacuated flasks and then was transferred to the flask in which the experiment was to be carried out (Figure 1). After solidification of the $C_2{}^3H_2$ (500 μCi) in the side-arm (d) by liquid nitrogen, the flask was opened and samples (a), (b) and (c) introduced. The experiment was started by sealing the flask and allowing gasification and expansion of the acetylene atmosphere under normal conditions and temperature of 28°C.

Simultaneous experiments to detect nitrogenase activity, using the acetylene reduction technique, showed that the bacteria associated with sugar-cane roots showed a reasonable enzymatic activity after incubation in N-free medium (300 μmol/hr/root segment).

The experiment was terminated after 30 min and samples were subsequently treated as follows:

1. Prefixation in 40% paraformaldehyde aqueous solution for 3 hr.
2. Washing in 0.125 *M*, pH 7.2 cacodylate buffer for 1 hr
3. Post-fixation with 2% osmium tetroxide in 0.25 *M* cacodylate buffer for 1.5 hr
4. Prestaining with 2% uranyl acetate in 75% acetone aqueous solution, for 16 hr
5. Gradual dehydration in increasing acetone concentrations and pure acetone for 1 hr, completing the process by treating with propylene oxide

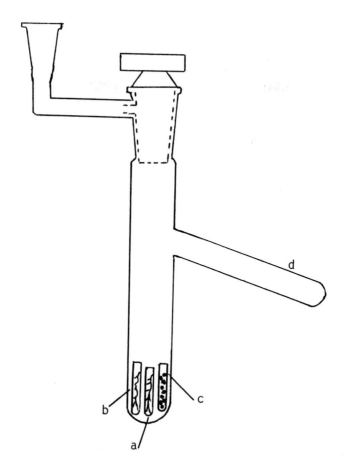

FIGURE 1. Experimental flask, with (a) field grown sugar-cane roots, (b) sugar-cane roots incubated for 24 hr in N-free medium, (c) *Phaseolus vulgaris* nodules as comparison/control, and (d) sidearm for solidified $C_2{}^3H_2$.

6. Soaking in 1:1 and 2:1 epon-propylene oxide at 37°C for 2 and 19 hr, respectively
7. Placed in pure epon at 4°C for 72 hr with subsequent polymerization at 37, 45 and 60°C for 18, 15, and 24 hr, respectively
8. Ultrathin sections of the different samples were placed on grids with parlodion and covered with Ilford L4 photographic emulsion according to Stevens' technique[8]
9. After 2.5 months in a dark chamber at 4°C they were developed with Microdol-X and fixed with 3% sodium thiosulphate aqueous solution
10. Finally stained with 2.5% uranyl acetate aqueous solution and lead citrate according to Reynolds[4]

III. RESULTS AND DISCUSSION

As previously found by Arias et al.,[1] examination of the ultrathin sections by electron microscopy revealed bacteria in the interior of the cell and in some fields bacteria

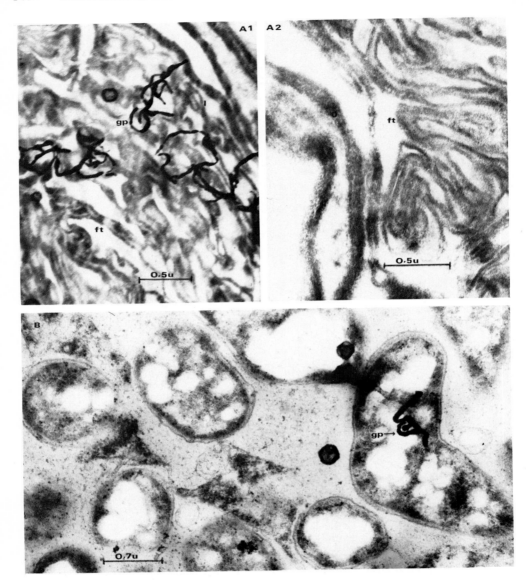

PLATE 1 Electron microradiograph showing: A1 — silver grains on "tubular growths" (ft), in cells of sugarcane roots incubated in nitrogen-free medium and treated with tritiated acetylene; A2 — improved view of "tubular growths" (ft), in cells; B — silver grains (gp) on bacteroids of nodules of bean root.

were noted passing through the cell wall. Also noted was the presence of silver grains located mainly on a structure made of twisted tubular formations with a diameter bigger than that of the endoplasmic reticulum (Plate 1 A). The presence of silver grains in the noninoculated roots was almost nil and less in the nodules (Plate 1 B) than in the roots preincubated in culture medium, indicating a very low background.

Counting of the silver grains seen in the electron-microradiographs on and outside the tubules, permitted the application of the χ^2 test, and the preparation of a 2 × 2

Table 1
COMPARISON BETWEEN THE NUMBER OF SILVER GRAINS ON AND OUTSIDE THE TUBULAR STRUCTURE, BY χ^2 TEST IN ROOTS PREINCUBATED IN CULTURE MEDIUM (24 HR)

Silver grains	Observed frequency	Expected frequency
On the tubular structure	42	27
Outside the tubular structure	13	28

Note: $\chi^2 = 16.37$. At a 0.1% probability level the difference is significant.

contingence table. Circles with a radius corresponding to 1250 Å, centered in the silver grains and covering an area with 50% probability of finding an intracellular radiation source, were used in this test. By this process, the grains produced by radiation originating from the tubular structure and those located outside it, could be detected more accurately. The number of the grains on the structure were taken as observed frequency.

The expected frequency on and outside the tubular structure was obtained by counting circles of the same radius drawn at random on a transparent plastic material which coincided with each of the mentioned cellular areas. The respective values were adjusted for a total of 55 circles corresponding to the total of silver grains found in the various electron-microradiographs. The results obtained are shown in Table 1.

The bean nodules showed a smaller number of silver grains and only from bacterial cells of *Rhizobium,* which probably indicates that in nodules not all the bacteria are active at the same time. It might however reflect the loss of unstable tritium labeled product, so that the method is underestimating the nitrogenase activity. The significant occurrence of silver grains in the tubular structures which were sometimes quite close to bacteria is difficult to interpret; however, they were found in roots previously incubated in N-free medium.

Although the possibility exists that the radioactive ethylene may have been bound by metallo-proteins not connected with N_2-fixation, the fact that with bean nodules silver grains are only connected with *Rhizobium* bacterial cells clearly suggests that they are in fact directly related to N_2-fixation. The fact that the "control" bean nodules showed a lower frequency of silver grains than incubated sugar-cane root, strongly indicates N_2-fixation within the latter.

From the structural point of view, the preliminary results also showed compartments with formations containing fibril material both in the vicinity and inside the cell wall. Growths similar to the nodule bacteroids, comprising electrolucent corpuscles, characterized as accumulation of beta-hydroxybutyric acid (Plates 2 A and 2 B), were noted in sugar-cane roots which had not been incubated in the culture medium.

Further research will be carried out to optimize the technique in order to obtain data which can be more accurately used in the interpretation of the results, although the main conclusion seems clear that nitrogenase activity, and hence N_2-fixation, can occur inside the root cells of sugar cane. Apart from its possible use in other associative N_2-fixing systems the method might have applications in the investigation of nodule systems to determine, e.g., the proportion and location of *Rhizobia* or other microorganisms actually engaged in N_2-fixation.

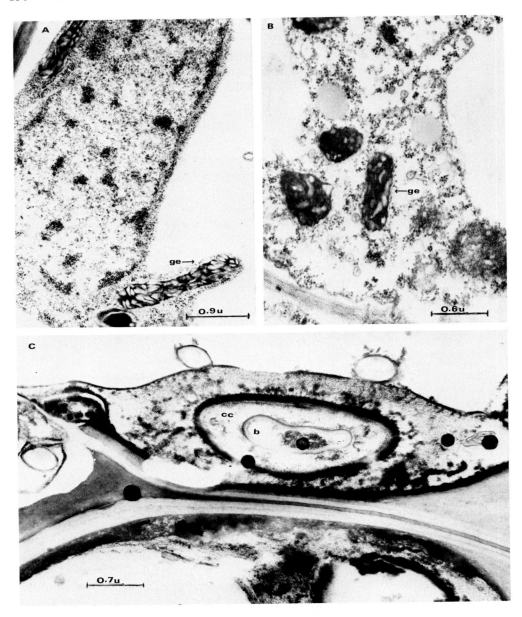

PLATE 2 Electron microradiographs showing: A — growths like bacteroids with electrolucent granules (ge) in the periphery, in sugarcane root cell nonincubated in culture medium; B — at one side, bacteroid of bean root nodule with similar granules; C — a bacteria (b) can be seen in the interior of a compartment in the cytoplasm (cc) of sugarcane root.

REFERENCES

1. **Arias, O. A., Gatti, I. M., Silva, D. M., Ruschel, A. P., and Vose, P. B.,** Primeras observaciones al microscopio eletrónico de bacterias fijadoras de N_2 en la raíz de la caña de azucar (*Saccharum* sp.), *Turrialba,* 28(3), 203, 1978.
2. **de-Polli, H., Matsui, E., Döbereiner, J., and Salati, E.,** Confirmation of nitrogen fixation in two tropical grasses by $^{15}N_2$ incorporation, *Soil Biol. Biochem.,* 9, 119, 1977.
3. **Döbereiner, J., Day, J. M., and Dart, P. P.,** Nitrogenase activity in the rhizosphere of sugar cane and some other tropical grasses, *Plant Soil,* 37, 191, 1972.
4. **Reynolds, E. S.,** *J. Cell Biol.,* 17, 208, 1963.
5. **Ruschel, A. P., Salati, E., and Henis, Y.,** Nitrogen-15 tracing of N-fixation with soil-grown sugar-cane seedlings, *Soil Biol. Biochem.,* 7, 181, 1975.
6. **Ruschel, A. P., Victoria, R. L., Salati, E., and Henis, Y.,** Nitrogen fixation in sugar cane (*Saccharum* sp.), *Ecol. Bull. Stockholm,* 26, 297, 1978.
7. **Ruschel, A. P. and Vose, P. B.,** Present Situation Concerning Studies on Associated N_2-Fixation in Sugar Cane, CENA Bol. Científico, BC-045, Centro de Energia Nuclear na Agricultura, Paracicaba, S. P., Brasil, 1977, 28.
8. **Stevens, A. R.,** *Methods in Cell Physiology,* Vol. 2, Prescott, D. M., Ed., Academic Press, New York, 1966, 255.

Chapter 21

USE OF ^{15}N ENRICHED GAS TO DETERMINE N_2-FIXATION BY UNDISTURBED SUGAR-CANE PLANT IN THE FIELD

E. Matsui, P. B. Vose,* N. S. Rodrigues, and A. P. Ruschel**

TABLE OF CONTENTS

*   Research supported by the International Atomic Energy Agency/United Nations Development Program
    and Comissão Nacional de Energia Nuclear.
** Centro de Energia Nuclear na Agricultura, Piracicaba, S. P., Brazil.

I. INTRODUCTION

Evidence of dinitrogen fixation by sugar cane was obtained by Ruschel et al.[2,5] and by De Polli et al.[1] for *Paspalum notatum* and *Digitaria decumbens,* using ^{15}N enriched atmosphere in a totally enclosed growth chamber in the laboratory.

The present work was carried out with the objective of developing a method to determine dinitrogen fixation by sugar cane under undisturbed field conditions, the soil atmosphere being continuously supplied with ^{15}N enriched nitrogen.

II. MATERIAL AND METHODS

A. Plant

An experiment with 11-months old "plant" cane was carried out in a sugar-cane field located at the Experimental Station of the Agronomic Institute of Campinas, 3 km distant from the Centro de Energia Nuclear na Agricultura. Ammonium sulphate at a rate of 40kg/ha had been applied in September 1978.

One plant with three stalks at different development stages was chosen, the plant separated a little from the others, to facilitate the setting up of the experiment. The experiment was carried out over a period of 5 days, 6 to 11 December 1978.

B. Preparation of the Gas

The dinitrogen ([15]N$_2$) was obtained from ammonium sulphate containing 90.3 atom % [15]N by reaction with lithium hypobromite, as is usually done in ^{15}N analysis by mass spectrometry. Figure 1 shows the system used to obtain, purify, measure the volume, and to adjust O$_2$ composition in the vessel used in the transport of the gas.

Initially the tank (R) was completely filled with water, including the tubes connecting the tank with valve (5) and with the water tank (W); 15 g of 90.3 atom % ammonium sulphate were placed each time in flask B through tube A, distilled water being used to carry the material to flask B. Closing valve 1, vacuum was applied to flask B and the tubes from B to valve 4, by using a mechanical vacuum pump, and opening valves 6, 2, and 3.

With valve 6 closed, filling tube A with the reagent which it passed on to B, tube A being always kept filled with hypobromite. When N$_2$ pressure in flask B reached atmospheric pressure, valves 2, 3, and 4 were opened, pump P was connected, and valve 5 opened to transfer the gas from B to R, passing through T where the gas was purified. Hypobromite was continuously added until all the ammonium sulphate had reacted, which could be noted by the solution in B becoming yellowish. When the reaction ended, flask B was slowly filled with distilled water; at the same time the circulation pump P transferred the gas to R. When the gas transfer was completed, valve 5 was closed, flask B was washed with distilled water and another 15 g ammonium sulphate added and the whole operation repeated. A total of 90 g ammonium sulphate was used.

About 18 ℓ N were obtained and transferred to tank R. Commercial O$_2$ was then added to reach 17.5% O$_2$. The gas thus prepared had the following molecular composition:

$$N_2 \; = \; 81.4\% \qquad\qquad CO_2 \; = \; 0.034\%$$

$$O^2 \; = \; 17.5\% \qquad\qquad NO \; = \; 0.009\%$$

$$Ar \; = \; 0.98\% \qquad\qquad N_2O \; = \; 0.02\%$$

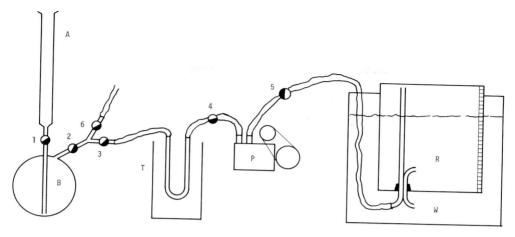

FIGURE 1. Diagram of the apparatus for $^{15}N_2$ atmosphere preparation and blending.

The isotopic composition of the nitrogen was now 86.7 atom % ^{15}N. Flask B containing the gas was ready for transport to the field when valve 5 was closed and the circulation pump disconnected.

C. Mounting of the System in the Field

A cylinder, 56 cm in diameter and 45 cm high, constructed from half an oil drum, was introduced into the soil around the chosen plant, the upper raised flange close to the stem being sealed. The cylinder was placed in position by hammering with a heavy hammer, the cylinder being protected with a wood block. A diagram of the mounted system is shown in Figure 2. A piece of automobile tire inner tube was used to provide a flexible seal. This was tied around both the cylinder flange and the stem with strips of rubber. The proprietary gas sealing compound "mastique elastico Igas" was then used to provide a gas tight seal.

Two air soil sampling tubes (C and D) were installed at 18 and 38 cm depth and two tubes on the soil surface (A and B), under the top of the cylinder. These tubes were used for the introduction of the gas, circulation of air through soda-lime to absorb CO_2, sampling of soil surface air and replenishment of O_2. The gas tank R was slowly introduced into the system through tube A. About 1 hr was spent on this operation. Gas composition and ^{15}N concentration in the soil surface, at 18 and 36 cm depth, respectively, were determined by periodic gas sampling in A, C, and D. The circulation pump P was periodically connected to circulate the gas through the soda-lime to take off CO_2 excess; oxygen was injected into the circulation line to replace that used.

There had previously been some doubts as to whether it was feasible to use a cylinder open at the bottom without serious loss of $^{15}N_2$ atmosphere. However, earlier tests of the system using argon, which has the advantage of being relatively cheap and analyzable directly in the mass spectrometer, had confirmed that there was negligible loss from diffusion. During the experiment some effect of temperature was noted, moving the gas up or down the profile, due to sun heating the top of the cylinder with subsequent cooling at night. In future this can be avoided by attaching a motor car tire inner tube as an expansion vessel.

At the conclusion of the experiment the plant was dissected into component parts: leaves, stem, stem base plus strut roots (*touceira*), and roots. The cylinder was excavated complete with all soil, and it was observed that only one or two roots had been cut by the cylinder and that none were visible deeper than 38 cm. The roots were

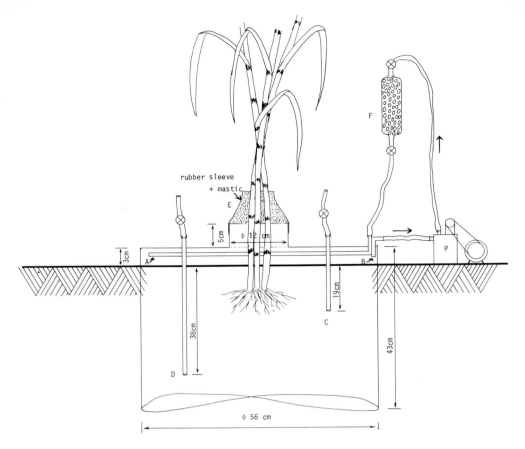

FIGURE 2.   Diagram of the system mounted in the field.

carefully separated from the soil, according to plow layer 0 to 19 cm, and below, 19 to 38 cm, corresponding to major differences in the occurrence and density of the roots. The soil of the two layers was weighed and sampled, and a further subsample of soil close to the roots was made.

III. RESULTS AND DISCUSSION

The weights and nitrogen content and ^{15}N analyses of the different plant parts are given in Table 1, the data for the soil system in Table 2.

In the plant system it is apparent that no excess ^{15}N can be detected except for the possibility that the stalk bases plus strut roots, *touceira*, may show slight enrichment, but clearly not statistically significant. In the soil system, the soil that was carefully separated from near the big roots (0 to 19 cm depth) showed enrichment as did a mixed sample of soil taken from small roots throughout the whole depth. Separate sampling of the soil from near large roots at 19 to 38 cm was not practicable, as the soil separated from the roots with excavation. No enrichment of bulk soil was observed. Due to the indeterminate nature of soil samples from the rooting zone, no quantification of fixation was possible.

Time variations of N_2, O_2, CO_2, and ^{15}N contents at 0, 19 and 38 cm depth are shown in Figures 3, 4, and 5. O_2 content of soil air outside the chamber at 19 cm depth

Table 1
YIELD OF DRY MATTER AND DISTRIBUTION OF N IN THE SYSTEM AT THE CONCLUSION OF THE EXPERIMENT

	Dry weight (g)	% N	N (mg)	atom % ^{15}N
1. Youngest stalk				
Bud	1.60	1.40	22.40	0.367
Leaf 6	—	—	—	—
5	2.22	1.54	34.19	0.368
4	3.03	1.40	42.42	0.367
3	2.53	1.40	35.42	0.368
2	2.00	1.58	31.60	0.368
1	3.05	0.80	24.40	0.371
Total for leaves	14.44	(av. 1.35)	190.43	(av. 0.368)
Stalk	14.82	0.56	83.0	0.367
2. Middle stalk				
Bud	3.29	1.40	46.06	0.368
Leaf 6	3.92	1.54	60.37	0.367
5	5.00	1.49	74.50	0.367
4	4.80	1.63	78.24	0.368
3	4.60	1.72	79.12	0.368
2	7.15	1.03	73.65	0.366
1	5.76	1.08	62.21	0.365
Total for leaves	34.50	(av. 1.41)	474.15	(av. 0.367)
Stalk	63.87	0.56	357.67	0.370
3. Oldest stalk				
Bud	7.24	1.49	107.88	0.368
Leaf 6	—	—	—	—
5	5.02	1.56	78.31	0.368
4	5.48	1.27	69.59	0.366
3	5.13	1.65	84.65	0.367
2	7.88	1.08	85.10	0.366
1	6.77	0.89	60.25	0.366
Total for leaves	37.51	(av. 1.32)	485.78	(av. 0.367)
Stalk	68.10	0.50	340.05	0.368
Stalk Bases *touceiro* + strut roots	118.20	0.39	461.00	0.371
Roots 0—19 cm	33.75	0.39	131.63	0.369
19—38 cm	10.90	0.36	39.27	0.368
Original "seed" stalk				0.368
Total for all leaves	86.48		1.150.36	44.8% of total
Total for 3 stalks	146.79		780.72	30.0% of total
Total for stalk bases	118.20		461.00	18.0% of total
Total for all roots	44.65		170.90	6.6% of total
				(99.7%)
Total for whole plant	369.12		2.569 g N	
Control: sugar cane from the same plot				0.368
^{15}N measurement, SE × 2				0.003 atom %

remained around 15%, varying between 14.3 and 16.5% and CO_2 content around 2.8%, varying between 2.0 and 3.2%. Figure 5 shows that decline in atom % ^{15}N in the soil atmosphere was linear with time which would have made it possible to apply isotope dilution theory, if excess ^{15}N had been determined in the leaves.

The inability on this occasion to demonstrate ^{15}N enrichment in the plant is possibly a little surprising, considering that there is demonstrable enrichment in the rhizosphere soil. The conclusion must be either that the fixed ^{15}N was largely retained within the

Table 2
WEIGHT OF SOIL SYSTEM, N CONTENT AND ATOM
% ¹⁵N AT THE CONCLUSION OF THE EXPERIMENT

	Dry weight Kg	Kjeldahl-N %	Total-N g	Atom % ¹⁵N
Soil, 0—19 cm	54	0.08	43	
19—38 cm	62	0.08	50	
Soil, sample from close to big roots, 0—19 cm depth				0.377
sample from close to small roots, 0—19 cm depth				0.372
Mixed sample from close to small roots, 0—38 cm depth				0.378
sample of mixed bulk soil				0.369
Control soil sample from outside container				0.369
¹⁵N measurements, SE × 2				0.003

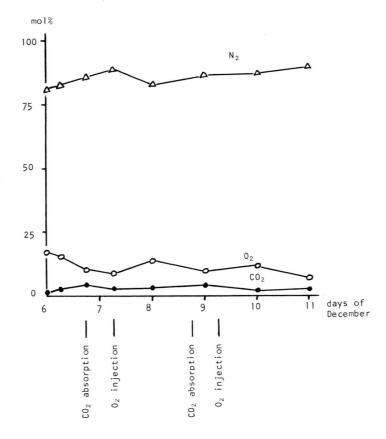

FIGURE 3.   Time course of variation in soil atmosphere composition in the top 0 to 19 cm deep zone.

fixing bacteria, or that we were unable to detect it in the leaves due to isotopic dilution, considering that there was a total of 2.57 g N in the plant. Retention by bacteria is supported by work on the uptake of ¹⁵N₂ by sugar cane in culture solution[3] where ¹⁵N-fixed is apparently held by the bacteria on or in the root, with a relatively small proportion immediately transferred to the leaves.

The difficulty of determining ¹⁵N excess in the leaves is compounded by the fact that with tillers of *Gramineae* it is impossible to predict at a given time which tiller or leaf

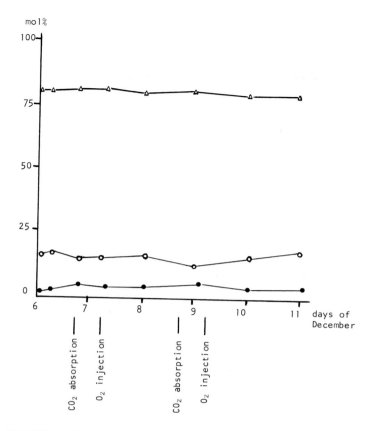

FIGURE 4.   Time course of variation in soil atmosphere composition in the 19 to 38 cm deep zone.

of a tiller is receiving preferential supply of nutrients or assimilates. However it is unlikely that any fixed ^{15}N absorbed by the plant was distributed uniformly. There is clearly some variation, though not significant, between the atom % ^{15}N content of different leaves, there being a tendency for the oldest leaves of the two oldest and biggest stalks to have the lowest ^{15}N content.

The question arises as to where all the ^{15}N$_2$ went. Some was left in the soil atmosphere at the end of the experiment, and some must have been lost by diffusion, though we believe this was relatively small. There was an apparently intense soil microbiological activity during the course of the experiment, as shown by the rapid rise in CO_2 concentration following successive efforts to remove excess. The balance of probabilities therefore suggests that much ^{15}N$_2$ was incorporated into microorganisms and became undetectable in the bulk soil due to the high dilution factor. Table 2 shows a total of about 93 g N to have been present in the soil system. The root system of sugar cane is very extensive and we were only able to separate a small part of the rhizosphere soil. Therefore, the total amount of ^{15}N$_2$ fixed close to the roots is likely to have been relatively high, but this has been masked by incorporation of the soil in the bulk.

Of course, the amount of potentially fixed N$_2$ which we attempted to measure was extremely small. If for purposes of illustration we assume that as much as 100 kg N might be fixed per hectare over 200 days, then this would be the equivalent of 0.5 kg/ ha/day or 2.5 kg/ha in 5 days. However, the root systems of 9-month old "plant" cane are not occupying the whole area, but are present in strips of soil a little less than

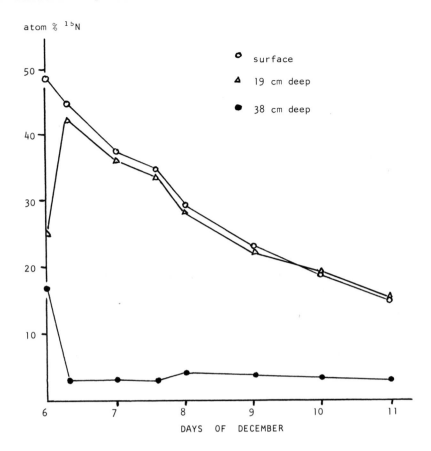

FIGURE 5.   Decrease in ^{15}N in the soil atmosphere with time.

56 cm wide. Sugar cane is planted in rows 1.4 m apart, so that in a hectare there will be 7142 m of rows 0.56 m wide, equivalent to 4000 m² occupied by roots.

This area will represent the area effectively available for associative N₂-fixation, so 2500 g N would need to be fixed during 5 days by 4000 m², equivalent to 0.625 g N/ m². But the container was 56 cm in diameter and covered an area of 2462 cm² or 0.2462 m². Therefore the amount that would need to be fixed within the container, to be equivalent to 100 kg N/ha in 200 days, is only 0.154 g N. If the amount fixed per hectare is less than 100 kg N annually, or if this amount is integrated over a longer growing period than 200 days, or if the period of active N₂-fixation is restricted to certain stages of plant growth, then the potential amount fixed during our rather arbitrary period of 5 days would be even less.

In retrospect, it is likely that we did not carry out the experiment during a period of maximum fixation rate. Acetylene analyses on field samples are very variable from time to time, and we have as yet no evidence as to the distribution of N₂-fixation during the 15 to 18 month growing period.[4] It could be that there are peaks of fixation corresponding to growth stages. The cane at 11-months old was entering a period of very rapid elongation following a dry period, and elongated substantially during the course of the experiment. It is likely that there would be little root exudation to stimulate a rhizosphere population. It should be noted too, that the weather conditions were unfavorable during the experiment. For approximately 70% of the period of the experiment the weather was overcast with rain of varying intensity, interspersed with periods of better weather, especially towards the end.

While the direct evidence for N_2-fixation associated with field grown sugar cane was shown in this work, further direct $^{15}N_2$ experiments with sugar cane must await an extensive survey program of regular acetylene reduction analyses throughout the growing period, in order to determine better the most intense periods of N_2-fixing activity in the field. The use of an open-ended cylinder for undisturbed studies was satisfactory and could be recommended for legume studies, where the N_2-fixing *Rhizobium* is in especially close relationship with the plant, and hence positive ^{15}N enrichment of leaf tissue will be more readily found.

Note added in proof: It has been subsequently found that there is a very rapid gas transfer via the shoots to the soil. When a similar experimental system was used and an acetylene atmosphere presented to the shoots within a polyethylene bag, acetylene was found in the rooting zone within 30 min. Gas transfer was greatly reduced with the onset of darkness, presumably through closure of the stomata. Our ^{15}N atmosphere must have therefore been diluted considerably during the course of the experiment.

ACKNOWLEDGMENT

The cooperation of the Estacão Experimental, Piracicaba (Instituto Agronomico, Campinas) in providing the experimental site and power supply is gratefully acknowledged.

REFERENCES

1. de Polli, H., Matsui, E., Döbereiner, J., and Salati, E., Confirmation of nitrogen fixation in two tropical grasses by $^{15}N_2$ incorporation, *Soil Biol. Biochem.*, 9, 119, 1979.
2. Ruschel, A. P., Henis, Y., and Salati, E., Nitrogen-15 tracing of N-fixation with soil-grown sugarcane seedlings, *Soil Biol. Biochem.*, 7, 181, 1975.
3. Ruschel, A. P., Matsui, E., Vose, P. B., and Salati, E., Potential N_2-fixation by sugar cane *Saccharum* spp. in solution culture. II. Effect of inoculation; and dinitrogen fixation as directly measured by $^{15}N_2$, *Associative N_2-Fixation*, Vose, Peter B. and Ruschel, A. P., Eds., CRC Press, Boca Raton, Fla., 1981.
4. Ruschel, A. P., Associative N_2-fixation by sugar cane, *Associative N_2-Fixation*, Vose, Peter B. and Ruschel, A. P., Eds., CRC Press, Boca Raton, Fla., 1981.
5. Ruschel, A. P., Victoria, R. L., Salati, E., and Henis, Y., Nitrogen fixation in sugar cane (*Saccharum officinarum*), environmental role of nitrogen-fixing blue-green algae and asymbiotic bacteria, *Ecol. Bull. Stockholm*, 26, 297, 1978.

Chapter 22

RELATIONSHIPS BETWEEN *AZOSPIRILLUM* SPP. ISOLATES FROM MAIZE AND SUGAR CANE

Siu Mui Tsai Saito* and L. A. Graciolli**

TABLE OF CONTENTS

*   Centro de Energia Nuclear na Agricultura, Piracicaba, S. P., Brazil.
** Faculty of Agronomy of Paraguacu Paulista (ESAPP), S. P., Brazil.

I. INTRODUCTION

Studies with *Azospirillum* spp. have been mostly concerned with their physiological behavior and number of experiments have shown the influence of mineral N, soil temperature and light intensity, for example, on N_2-fixation activities of the bacteria.[1,9,2] The effect of inoculation on the increase of N in plants was beneficial in some cases[13] or not significant in others.[2,4]

Azospirillum spp. were isolated from *Zea mays, Saccharum* spp., *Sorghum aestivum, Digitaria decumbens, Pennisetum americanum, Panicum maximum* and *Oriza sativa*. In these cases, the species was not exhaustively classified. The classification of the *Azospirillum* species follows the physiological characteristics of the bacteria and determined as Groups I and II and, later on, Group III, but little is known as to the relationships between isolates from different plant species.

The occurrence of *Azospirillum* species in the rhizosphere or roots of many tropical and some temperate grasses has been widely observed and studied. In many cases, a general incidence of the bacteria was observed, but a reexamination of its ecological distribution will be needed as the preference of bacteria to establish in one specific host has been demonstrated, as in the case of *Azotobacter paspali* with *Paspalum notatum*[7] and some species of *Bacillus* with wheat.[8]

The purpose of this work was to make a preliminary survey of the serological characteristics of different isolates of *Azospirillum* species from maize and sugar cane, when compared with a known and identified isolate of *Azospirillum* from *Paspalum* roots.

II. MATERIAL AND METHODS

A. Identification and Antiserum Preparation of SCPA

The *Azospirillum* test line isolated from *Paspalum* roots (SCPA) was identified as *A. brasilense* nir⁻.[12,14] The bacterium was inoculated in flasks containing liquid N-free plus malate medium[6] and incubated at 37°C over a week. The cells were washed three times with sterile saline (0.85% NaCl) at 10,000 r/min for 10 min and resuspended in sterile saline to give a concentration of around 10^8 to 10^9 cell/mℓ. The method of inoculation in the rabbit was described by Titova,[15] the injection sequence of which is shown in Table 1. After a period of 2 months of treatment, the blood was collected and the antiserum was separated by centrifuge, giving a titre of about 1:3200. Reactions with the original antigen gave extremely clear precipitation followed by abundant agglutination in the test tubes until a dilution of 1:3200.

B. Isolation of *Azospirillum* lines

Whole plants of sugar cane and maize were collected from Araras Experimental Station (Araras, São Paulo) and from Campinas University Experimental Station (Campinas, São Paulo), respectively. Pieces of plants, roots and leaves, were surface sterilized with 20% HClO₄ for 3 min and washed with sterile water several times. Soil from rhizosphere of sugar cane was diluted in sterile saline. One milliliter of each extract (until dilutions 10^6) and plant pieces were incubated in tubes containing a modified semisolid NFb media and nitrogenase activity was measured to select positive tubes.[10] One loopful of each tube was streaked out on semisolid NFb medium plus 0.002% yeast extract to observe the pellicle formation. Isolation of pure colonies was obtained by streaking inoculum on potato infusion solid medium[5] and a typical pink and wrinkled colony of each plate was isolated and inoculated in solid NFb media for serological tests.

Table 1
SCHEME OF SCPA INOCULATION IN RABBIT

Injection	Volume antigen (ml)	Protein (mg)	Method
1st	0.5 Adjuvant + 0.5 Antigen	10	Subcutaneous, 7 days interval
2nd	0.5	10	Intravenous, 7 days interval
3rd	0.5	10	Subcutaneous
	0.5	10	Intravenous (2 hr later) (Take subsample at 8th and 10th day)
4th	0.5	10	Subcutaneous
	0.5	10	Intravenous (2 hr later)
5th	0.5	10	Subcutaneous
	1.0—2.0	20—40	Intravenous (2 hr later)
6th	1.0	20	Subcutaneous
	3.0	60	Intravenous (2 hr later) (Take subsample at 8th and 10th day)

Modified from Titova, E. V., *Entomol. Rev.,* 49 155, 1970. With permission.

C. Characterization of the Isolates

All isolates were tested in three media: (1) semisolid NFb, (2) semisolid NFb plus 0.5% glucose without malate and indicator, and (3) semisolid NFb plus 5 mM NH₄NO₃. Microscope observations and the catalase test were also included thus making it possible to distinguish *A. lipoferum* from *A. brasilense,* and nir⁺ from nir⁻ from both species.

D. Serological Tests

Each *Azospirillum* isolate was tested against SCPA antiserum. For this reaction, the isolates were harvested from the solid NFb media, resuspended in saline to give 10^8 to 10^9 cells/ml. Of this antigen 0.05 ml of each isolate was used to react with 0.5 ml of the SCPA antiserum (diluted to 1:50). Two sets of reaction were prepared: one set was heat treated (boiled for 30 min) and the other was used normally, to give somatic (0 antigen) and surface reactions (H antigens), respectively. Controls using only SCPA antigen plus saline were used in both cases.

III. RESULTS AND DISCUSSION

Table 2 shows the identification of the *Azospirillum* isolates according to Sampaio and Döbereiner[12] and Tarrand et al.[14] The serological reactions are shown in Table 3. Most isolates from maize roots were identified as *A. lipoferum* nir⁻ and from sugar cane as *A. brasilense* nir⁻, confirming the work of Baldani and Döbereiner,[3] which has demonstrated mostly the presence of *A. lipoferum* in maize roots and of *A. brasilense* nir⁻ in wheat and rice roots (Table 2). Also Rocha et al.[11] found the preferential establishment of *A. brasilense* in sugar cane roots.

A. lipoferum isolates from maize did not interact with antiserum of SCPA, identified as *A. brasilense* nir⁻, but in sugar cane a higher serological affinity was observed, though not as strong as the specific reaction of SCPA (Table 3). Using whole and unboiled cells, mostly agglutination was observed in the positive tubes. When the cells were heat treated, the reactions were more in the form of precipitation. These results

Table 2

ORIGIN AND IDENTIFICATION OF *AZOSPIRILLUM* SPP. ISOLATES FROM THEIR CHARACTERISTIC GROWTH IN DIFFERENT MEDIA, MICROSCOPIC OBSERVATIONS AND CATALASE REACTIONS

Isolate number	Origin	Growth			Alkaline medium	Catalase	Identification
		NFb	NFb* + glu	NFb + NH₄NO₃ (bubbles)			
AMR-1	Maize roots	Good	Good	A few	Large polymorph	Negative	*A. lipoferum* nir⁺
AMR-2	Maize roots	Good	Good	No	Large polymorph	Negative	*A. lipoferum* nir⁻
AMR-3	Maize roots	Good	Good	No	Large polymorph	Negative	*A. lipoferum* nir⁻
AMR-4	Maize roots	Good	Good	No	Large polymorph	Negative	*A. lipoferum* nir⁻
AMR-5	Maize roots	Good	Poor	No	Normal, v. motile	Positive	*A. brasilense* nir⁻
ASR-1	Sugar cane roots	Good	Good	No	Large polymorph	Negative	*A. lipoferum* nir⁻
ASR-2	Sugar cane roots	Good	Poor	No	Normal, v. motile	Positive	*A. brasilense* nir⁻
ASR-3	Sugar cane roots	Good	Good	No	Large polymorph	Negative	*A. lipoferum* nir⁻
ASR-4	Sugar cane roots	Good	Poor	No	Normal, v. motile	Positive	*A. brasilense* nir⁻
ASR-5	Sugar cane roots	Good	Poor	No	Normal, v. motile	Positive	*A. brasilense* nir⁻
ASR-6	Sugar cane roots	Good	Poor	No	Normal, v. motile	Positive	*A. brasilense* nir⁻
ASS-1	Sugar cane rhizosphere	Good	Poor	No	Normal, v. motile	Positive	*A. brasilense* nir⁻
ASS-2	Sugar cane rhizosphere	Good	Poor	No	Normal, v. motile	Positive	*A. brasilense* nir⁻
ASL-1	Sugar cane leaves	Good	Poor	No	Normal, v. motile	Positive	*A. brasilense* nir⁻
ASL-2	Sugar cane leaves	Good	Poor	No	Normal, v. motile	Positive	*A. brasilense* nir⁻
SCPA	*Paspalum* roots	Good	Poor	No	Normal, v. motile	Positive	*A. brasilense* nir⁻

[a] Without malate and indicator.

Table 3
SEROLOGICAL TESTS USING SCPA ANTISERUM AGAINST ANTIGENS OF DIFFERENT *AZOSPIRILLUM* ISOLATES FROM SUGAR CANE AND MAIZE

Isolates	Identication	Serological tests	
		Agglutination (H)	Precipitation (O)
AMR-1	*A. lipoferum* nir +	−[a]	−
AMR-2	*A. lipoferum* nir⁻	−	−
AMR-3	*A. lipoferum* nir⁻	−	−
AMR-4	*A. lipoferum* nir⁻	−	−
AMR-5	*A. brasilense* nir⁻	−	+ +
ASR-1	*A. lipoferum* nir⁻	−	+
ASR-2	*A. brasilense* nir⁻	+ + +	+ +
ASR-3	*A. lipoferum* nir⁻	−	−
ASR-4	*A. brasilense* nir⁻	+ + + +	+ +
ASR-5	*A. brasilense* nir⁻	+	−
ASR-6	*A. brasilense* nir⁻	+ + +	−
ASS-1	*A. brasilense* nir⁻	+ + +	+ + +
ASS-2	*A. brasilense* nir⁻	+ + +	+ + +
ASL-1	*A. brasilense* nir⁻	−	−
ASL-2	*A. brasilense* nir⁻	−	+
SCPA	*A. brasilense* nir⁻	+ + + +	+ + + +

[a] Reaction: − = negative; + = weak; + + = moderate; + + + = strong; + + + + = very strong.

suggest that *A. brasilense* nir⁻ from sugar cane are more related to a *A. brasilense* nir⁻ isolated from *Paspalum,* this being mainly seen as surface interaction. Also, a weak but positive reaction (precipitation) was observed when *A. brasilense* nir⁻ from maize roots (AMR-5) was tested.

In sugar cane, both isolates of *A. lipoferum* did not show any reaction, but with other *A. brasilense* nir⁻ there were five positive reactions in a total of six isolates (four from surface sterilized roots and two from the rhizosphere). Isolates from sugar-cane leaves did not show reactions. Less intensity appeared when the same isolates were heat treated, with the exception of rhizosphere isolates when both treatments gave similar and positive reactions, denoting a higher affinity of SCPA with isolates from the rhizosphere than from inside the roots or leaves of sugar cane. An opposite result was obtained when testing *A. brasilense* nir⁻ isolates from sugar-cane leaves, which did not show the expected reactions.

All these results will need to be confirmed by further experiments, but at the present time they suggest that (1) *A. lipoferum* isolates from maize do not have serological interactions with a *A. brasilense* nir⁻ from *Paspalum* (SCPA); (2) it seems that the *A. brasilense* nir⁻ isolated from sugar cane are related to SCPA, mainly when the isolates are from the rhizosphere; and (3) there is evidence of surface interaction among those isolates from sugar cane with SCPA.

REFERENCES

1. Abrantes, G. T. V., Day, J. M., and Döbereiner, J., Métodos para estudo da atividade de nitrogenase em ráizes de gramineas colhidas no campo. Campinas, 15th Congr. Brasileiro Ci. Solo, *Bull. Int. Inf. Biol. Sol Lyon*, 21, 1, 1975.

2. Albrecht, S. L., Okon, Y., and Burris, R. H., Effects of light and temperature on the association between *Zea mays* and *Spirillum lipoferum, Plant Physiol.*, 60, 528, 1977.

3. Baldani, V. L. and Döbereiner, J., Host Plant Specificity in the Infection of Maize, Wheat and Rice with *Azospi̇rillum* spp., Int. Workshop Associative N$_2$-Fixation, Centro de Energia Nuclear na Agricultura, Piracicaba, S. P., Brasil, 1979.

4. Barber, L. E., Tjepkema, J. D., and Evans, H. J., Nitrogen fixation in the root environment of some grasses and other plants in Oregon, (Int. Symp. Environ. Role of Nitrogen-Fixing Blue-Green Algae and Asymbiotic Bacteria, Uppsala, Sweden) *Ecol. Bull. Stockholm*, 26, 366, 1978.

5. von Bülow, J. F. W. and Döbereiner, J., Potential for nitrogen fixation in maize genotypes, in Brasil, *Proc. Nat. Acad. Sci. U.S.A.*, 72, 2389, 1975.

6. Day, J. M. and Döbereiner, J., Physiological aspects of N$_2$-fixation by a *Spirillum* from *Digitaria* roots, *Soil Biol. Biochem.*, 3, 45, 1976.

7. Döbereiner, J., *Azotobacter paspali* sp. n, uma bactéria fixadora de nitrogênio na rizosfera de *Paspalum, Pesqui. Agropecu. Bras.*, 1, 357, 1966.

8. Larson, R. I. and Neal, J. L., Jr., Selective colonization of the rhizosphere of wheat by nitrogen-fixing bacteria, (Int. Symp. Environ. Role of Nitrogen-Fixing Blue-Green Algae and Asymbiotic Bacteria, uppsala, Sweden,) *Ecol. Bull. Stockholm*, 26, 331, 1978.

9. Nayak, D. N. and Rao, R. V., Nitrogen fixation by *Spirillum* sp. from rice roots, *Arch. Microbiol.*, 115, 359, 1977.

10. Patriquin, D. G., Graciolli, L. A., and Ruschel, A. P., Nitrogenase Activity and Bacterial Infection of Roots and Germinated Cuttings, Int. Symp. Nitrogen Fixation, Steenbock-Kettering Institute, Madison, Wis., 1978.

11. Rocha, R. E. M., Baldani, J. I., and Döbereiner, J., Host Plant Specificity in the Infection of C$_4$ Plants with *Azospirillum* spp., Int. Workshop Association N$_2$-Fixation, centro de Energia Nuclear na Agricultura, Piracicaba, S.P., Brasil, 1979.

12. Sampaio, M. J. A., Vasconcelos, L., and Döbereiner, J., Identification of three groups within *Spirillum lipoferum*, (Int. Symp. Environ. Role of Nitrogen-fixing Blue-green Algae and Asymbiotic Bacteria, Uppsala, Sweden), *Ecol. Bull. Stockholm*, 26, 364, 1978.

13. Smith, R. L., Schank, S. C., Bouton, J. H., and Quesenberry, K. H., Yield increases of tropical grasses after inoculation with *Spirillum lipoferum*, (Int. Symp. Environ. Role of Nitrogen-fixing Blue-green Algae and Asymbiotic Bacteria, Uppsala, Sweden), *Ecol. Bull. Stockholm*, 26, 380, 1978.

14. Tarrand, J. J., Krieg, N. R., and Döbereiner, J., A taxonomic study of the *Spirillum lipoferum* group, with descriptions of a new genus *Azospirillum* gen. nov. and *Azospirillum brasilense* sp. nov., *Can. J. Microbiol.*, 24, 967, 1978.

15. Titova, E. V., Use of the precipitation test in a study of inter-relationships between *Eurycaster integriceps* Put. (*Heteropter, Scutelleridae*) and predatory arthropods, *Entomol. Rev. USSR*, 49, 155, 1970.

Chapter 23

INOCULATION OF SOYBEANS (*GLYCINE MAX* L. MERRILL) WITH *RHIZOBIUM* JAPONICUM* AND INTERCROPPED WITH SUGAR CANE (*SACCHARUM* SPP.) IN VALLE DEL CAUCA, COLOMBIA**

D. J. Belalcazar Gutierrez***

TABLE OF CONTENTS

*   Cooperative work — Facultad de Ciencias Agrarias, CIAT, Ingenio Central de Castilla and Instituto Columbiano and Agropecuario.

** Translators: A. P. Ruschel and D. Athie.

***Facultad de Ciencias Agrarias, Universidad Nacional de Columbia, Palmira.

I. INTRODUCTION

Double-cropping is an alternative to increasing yield or area. Besides the possibility of cultivating two different crops in the same area,[1,7] there is the occurrence of symbiotic nitrogen fixation,[4,9] decrease in the need for N-fertilizer application,[6] more reduced population of *Empoasca Kraemeri* and *Crisomelidos* in the double crops than in monocrops[3] and some other beneficial effects on the soil.

Due to high protein content (more than 36%) and its ability to fix N₂ in symbiosis with *Rhizobium japonicum*,[2] soybeans have great potential for intercropping with sugar cane. The main objectives of the present work were (1) to select *R. japonicum* strains for better efficiency and quantification of nitrogen fixation; (2) to study different behavior of strains (interaction bacteria/genotype) according to the lines used; and (3) to compare soybean production with and without inoculation and its effect on sugar cane production and yield.

II. MATERIALS AND METHODS

Soybean varieties 116 and 106 (ICA Tunia) from the Legume Program of the Instituto Colombiano Agropecuario (ICA) were used. Growth period was 100 to 110 days, flowering at 35 to 50 days. CIAT 4, CIAT 51, and CIAT 90 *Rhizobium* strains were used. Sugar cane variety was CP 57-603. Experimental design was randomized blocks (12 × 12 m) with nine treatments and four replications. Treatments comprised three inoculations, one control, and one sugar cane monoculture. Used were 2.5 g inoculum per kilogram of seed. Inoculum used proved to be viable (1×10^8) in accordance with Vincent.[8]

Sugar cane was planted in rows 1.5 m apart and the soybeans were sown on the following day, 30 cm from the rows with 5 cm between each plant. No fertilizer was applied.

To estimate soybean-sugar cane competition, sugar-cane samples were taken to determine population (stalks/meter), plant height in the three central rows, at 60, 90, 104, and 153 days from planting. Production and yield were determined at harvesting. Acetylene reduction, nodule dry weight, and total nitrogen in the plant tops were determined (10 plants/block) at 43, 60, and 76 days from planting. Yield was determined at harvesting (102 days from planting).

III. RESULTS AND DISCUSSION

Table 1* shows acetylene reduction data and it is noticed that the highest nitrogenase activity was at 60 days, i.e., at the beginning of pod filling when variability is relatively small. Differences between treatments were analyzed during this period. The data at 60 days are presented in more detail and show greater acetylene reduction with var. 116; strains CIAT 4 and 90 being more promising when compared with var. 106 (ICA, Tunia). This indicates a greater interaction between the strains and genotype L-116, and there was a significant difference between genotypes.

Table 2* shows that there was no consistency between nodule dry weight and acetylene reduction, or when comparing nodule dry weight with total nitrogen of plant tops in line 116. Dry weight of nodules was higher in strain 51 and total nitrogen of the tops was best in strain 90. The behavior of strains in line 106 (ICA Tunia) was unexpected, since there was no correlation between the two variables. It was noted that in

---

* Analysis of variance available from author: Editors.

Table 1
ACETYLENE REDUCTION ACTIVITY AT DIFFERENT TIMES DURING THE GROWING PERIOD, EXPRESSED AS μMOL C_2H_4/PLANT/HR MEAN OF FOUR REPLICATIONS

		Days from sowing			
			60		
	43				76
Treatments		Mean of treatments	Mean of inoculated	Mean of lines	
L 116 noninoculated	8.12	22.78			2.51
L 116-strain CIAT 4	1.00	31.80			4.42
L 116-strain CIAT 51	3.91	27.20	29.43	27.77	5.20
L 116-strain CIAT 90	3.40	29.30			2.49
L 106 noninoculated	1.20	22.19			2.37
L 106-strain CIAT 4	0.90	12.58			4.16
L 106-strain CIAT 51	3.92	14.78	16.40	17.85	3.14
L 106-strain CIAT 90	11.40	21.84			2.65
Mean	4.22	22.81			3.37
S	3.51	12.57			2.82
CV	83.30	55.08			83.74

Table 2
DRY WEIGHT OF NODULES — GRAM PER TEN PLANTS — AND TOTAL NITROGEN IN THE PLANT TOPS — MILLIGRAM PER PLANT—AT 60 DAYS FROM SOWING

	Line			
	L 116		L 106	
Strain	Dry weight of nodules	Total nitrogen in plant tops	Dry weight of nodules	Total nitrogen in plant tops
Control	2.26	33.76	1.50	32.47
C-4	2.17	41.79	1.04	38.58
C-51	4.16	47.29	0.83	36.85
C-90	3.31	52.81	1.83	30.70
Inoculated-mean	3.21	47.30	1.23	35.38
Linear mean	2.98	43.91	1.30	34.65
Mean		2.14	39.28	
CV %		56.07	23.86	
DMS 5%		1.76	13.77	

L-116 there was a tendency for total N of the tops, and nodule dry weight, to be greater in the inoculated treatments than in the control. Generally there is an association between the variables, the correlation coefficient being 0.85 although the behavior of each genotype seems to be different (Figure 1). Association is direct in L-116 and indirect in L-106.

It is important to record the excellent nodulation of the noninoculated treatments, due to native populations. The different behavior of the strains in the different lines suggests host specificity for bacterial efficiency, and in the present case there was greater interaction between the bacteria and L-116.

Nitrogen fixation in each treatment was determined by acetylene reduction (Table

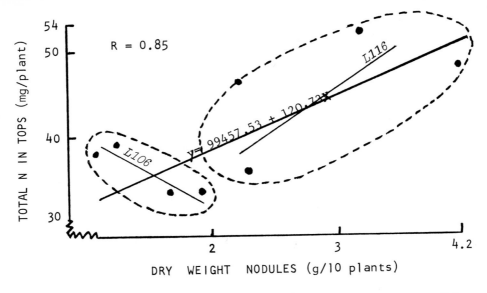

FIGURE 1.   Relationship between dry weight nodules and total N in the plant tops at 60 days from sowing.

Table 3
NITROGEN FIXATION — KILOGRAMS PER HECTARE — IN THE PERIOD 43 TO 76 DAYS FROM SOWING

Lines	Noninoculated	Strains CIAT 4	51	90	Mean Inoculated	Lines
L 116	60.2	98.6	78.1	90.8	89.2	81.9
L 106	70.3	34.6	38.8	51.1	41.5	48.7

3). Native strains in L-106 exceeded the inoculated by 28.8 kg N/ha, while in L-116 the inoculated exceeded the native strains by 29.0 kg N/ha. This proves bacterial/genotype interaction, the most effective strains being CIAT 4 and 90 in L-116.

Soybean results (Table 4) were highly satisfactory as 1.85 ton/ha were obtained, while commercial production as a monocrop in Valle del Cauca of the lines used is 2.5 to 2.8 t/ha. Yield obtained was therefore approximately 70% in 70% of the area, indicating that the soybean was not affected by the cane competition. Analysis of variance* showed no significant differences between treatments, which indicates that soil inoculation is not necessary for soybean production on soils similar to the one studied.

Table 5 shows the sugar-cane population and height, and a significant effect of soybean intercrop on sugar-cane population (stalks/m). Generally, there is an increment of sugar-cane height immediately after the soybean harvest which was 18.5 cm smaller than cane in monoculture at 153 days. Table 6 shows a recovery in population and height when comparing monoculture yield of 157 ton/ha with intercrop yield of 152 ton/ha. Treatment L-106 (IAC Tunia)-CIAT 51 was higher in cane yield when compared to monoculture. Sugar yield is shown in Table 7 and it can be noted that there was no significant difference between treatments with L-106 and monoculture yield, while it is better in monoculture than with L-116.

* Analysis of variance available from author: Editors.

Table 4
YIELD OF SOYBEANS PER HECTARE

Treatment	Mean of four replications (ton/ha)
1. L 116 noninoculated	1.90
2. L 116 — strain CIAT 4	1.81
3. L 116 — strain CIAT 51	1.83
4. L 116 — strain CIAT 90	1.80
5. L 106 noninoculated	1.95
6. L 106 — strain CIAT 4	1.74
7. L 106 — strain CIAT 51	1.90
8. L 106 — strain CIAT 90	1.90

Comparison	Treatments	Mean (ton/ha)	
L 116 vs. L 106	1, 2, 3, 4 vs. 5, 6, 7, 8	1.83	1.87
Inoculated vs. control/L 116	2, 3, 4 vs. 1	1.81	1.90
Inoculated vs. control/L 106	6, 7, 9 vs. 5	1.84	1.95

Table 5
HEIGHT AND POPULATION (STALKS PER METER ROW) OF SUGAR CANE IN MONOCULTURE AND INTERCROPPED

Variables	Type of culture	Time (days)			
		60	90	104	153
Height (cm)	Monoculture	20.6	31.7	35.0	67.2
	Intercropped	22.7	32.6	34.2	48.7
Population	Monoculture	10.3	20.3	19.3	22.6
	Intercropped	6.4	6.5	9.5	14.7

IV. CONCLUSIONS

1. Maximum nitrogen fixation occurred at the beginning of pod filling (60 days from sowing).
2. There was a marked decline in nitrogen fixation at the end of pod filling, possibly due to mobilization of nutrients and competition for photosynthesis products.
3. The decrease in fixation affects the ability of the plant to incorporate the nitrogen necessary for grain development which suggests that foliar fertilization at the end of pod filling might have a favorable effect in soybean production.
4. Line 116 showed the best nitrogen fixation.
5. The best interaction was between Line 116 and strains CIAT 4 and 90, this having the best nitrogen fixation.
6. Native populations of *Rhizobium japonicum* are relatively effective compared with inoculated strains.
7. The maximum amount of N_2 fixed was 98.6 kg/ha in the case of Line 116-Strain 4, and 60.2 kg/ha without inoculation.
8. With inoculation in Line 116 there was an increment of 29 kg N/ha which, related to 46% N-urea, represents U.S. $31.
9. Soybean production was highly satisfactory, with an average of 1.85 ton/ha. This represents approximately 70% of commercial production, however, it took only 70% of the area in the sugar cane-soybean system.

Table 6

YIELD OF SUGAR CANE[a] (TONS CANE PER HECTARE) IN MONOCULTURE AND INTERCROPPED WITH SOYBEANS

Treatment	Replications				Mean
	I	II	III	IV	
Monoculture control	186.83	151.45	153.85	139.29	157.85
Cane L 116 noninoculated	168.93	159.75	135.32	160.47	156.11
Cane L 116 CIAT 4	129.63	161.13	139.64	145.34	143.93
Cane L 116 CIAT 51	151.07	144.92	149.70	137.48	145.79
Cane L 116 CIAT 90	155.57	152.28	147.46	166.09	155.35
Cane ICA Tunia noninoculated	160.03	151.09	165.43	149.45	156.50
Cane ICA Tunia CIAT 4	146.54	156.39	147.29	141.81	148.00
Cane ICA Tunia CIAT 51	166.60	133.57	160.91	176.18	159.31
Cane ICA Tunia CIAT 90	158.78	159.39	143.20	149.04	152.60

[a] Cane variety CP 57-603.

Table 7

YIELD OF SUGAR[a] (TONS PER HECTARE) IN MONOCULTURE AND INTERCROPPED WITH SOYBEANS

Treatment	Replications				Mean
	I	II	III	IV	
Monoculture control	16.22	15.71	16.46	14.44	15.70
Cane L 116 noninoculated	15.24	14.78	13.13	15.77	14.73
Cane L 116 CIAT 4	13.64	13.78	14.37	14.90	14.17
Cane L 116 CIAT 51	15.65	15.64	12.92	11.05	13.81
Cane L 116 CIAT 90	15.09	15.96	14.19	14.25	14.87
Cane ICA Tunia noninoculated	17.14	14.07	17.52	14.42	15.78
Cane ICA Tunia CIAT 4	14.99	15.25	16.51	13.96	15.17
Cane ICA Tunia CIAT 51	16.49	13.69	16.73	14.57	15.37
Cane ICA Tunia CIAT 90	16.01	15.24	13.95	15.19	15.09

[a] Based on theoretical conversion.

10. From a production view-point, it is not necessary to inoculate normal soils in Valle del Cauca as they are rich in nitrogen and the native populations of *Rhizobium japonicum* are as effective as the inoculated strains.

11. Population and height of sugar cane do not affect the development of the soybean due to its earliness.

12. Soybean variety L-116 (ICA Tunia) showed the best behavior in relation to production and yield of sugar cane.

13. Sugar cane production and yield are relatively little affected by intercropping with soybean, achieving an average of 156 ton/ha and 15 ton of sugar per hectare, respectively.

14. Although production of cane was 5 ton/ha less (U.S. $50) in intercrop with soybean than in monoculture, 1.85 ton soybean per hectare, worth U.S. $575, can be produced in the same area.

REFERENCES

1. **Francis, C. A., Flor, C. A., and Temple, S. R.,** Selección de Variedades para Sistemas de Cultivo Intercalado en los Trópicos, presentado en el simposio sobre cultivos múltiples, Sociedad Americana de Agronomía (A.S.A.), Reunión Anual, Knoxville, Tenn., October 27, 1975.

2. **Freire, J. R. J.,** Inoculation of Soybeans, presented at the Workshop on Exploiting the Legume *Rhizobium* Symbiosis in Tropical Agriculture, Kahului, Manui, Hawaii, 1976.

3. **Garcia, J. E., Cardona, C., and Raigosa, J.,** Evaluación de poblaciones de insectos plagas en la asociación caña de azúcar — Frijol y su relación con rendimientos, CIAT Apartado aéreo 6713 — Cali; Ingenio Providencia, Apartado aéreo 224, Palmira, Colombia, 1979.

4. **Merriel, I. E. and Döbereiner, J.,** Potencial de Fixacao de N_2 e Incorporacao de N Mineral na Soja, EMBRAPA, Simp. Int. Sobre Limitacces e Potenciais da Fixacao Simbiotica de Nitrogenio nos Trópicos, U. Nal. de Brasilia Julho, Brasil, 1977.

5. Ministerio de Agricultura de Colombia — Programas Agrícolas, 63, 1977.

6. **Nutman, P. S.,** Perspectives in biological nitrogen fixation, *Sci. Prog., New Haven,* 59, 55, 1971.

7. **Pan, Y. C. and Lee, K. M.,** A study on the interplanting with soybean, *Taiwan Sugar Exp. Stn. Annu. Rep.,* 17, 228, 1963.

8. **Petroso, T. M. H. and Freire, J. R. J.,** Controle da qualidade dos inoculantes comerciais para soja, Trabalho apresentado pa *VIII Reuniao Latino Americana de Rhizobium* CIAT, Cali, Colombia, 1976, 68.

9. **Vincent, J. M.,** *Manual Práctico de Rhizobiología,* Editorial Hemisferio Sur., Buenos Aires, 1975.

Chapter 24

NITROGEN-FIXING POPULATION AND ACTIVITY ASSOCIATED WITH WETLAND RICE

Iwao Watanabe, Osamu Ito, and Wilfredo Barraquio*

TABLE OF CONTENTS

*   Soil Microbiology Department, International Rice Research Institute, Los Banos, Laguna, Philipines.

I. INTRODUCTION

Sen[5] drew attention to the presence of N_2-fixing bacteria in the surface-sterilized roots of wetland rice. This finding had been long overlooked until Yoshida and Ancajas[11] and Balandreau et al.[1] found that roots of wetland rice have the ability to fix N_2 due to the heterotrophic bacteria that live on and in the rice roots. This ability was detected by acetylene reduction technique. It was later found that acetylene-reducing activity of the excised roots of wetland rice was much higher than that of dryland rice.[12] Therefore, the possible contribution of rhizospheric nitrogen fixation was suggested to explain the maintenance of nitrogen fertility in flooded rice soils without nitrogen fertilizer.

This paper reports some of the research at the International Rice Research Institute (IRRI) on "rhizospheric nitrogen fixation" in wetland rice.

II. EVIDENCE OF N_2-FIXATION ASSOCIATED WITH RICE CONFIRMED BY $^{15}N_2$ INCORPORATION

In recent studies at IRRI[4,10] acetylene reduction technique was used to estimate N_2-fixing activity associated with the rice plant. To determine directly the N_2-fixing activity associated with the rice plant and the translocation of the fixed nitrogen, an experiment using ^{15}N tracer method was conducted. The plants at heading stage were taken from the field and water-cultured in gas-tight growth chamber, then fed with $^{15}N_2$ gas. The chamber had provision for gas mixture circulation, for removal of moisture, and for maintenance of carbon dioxide concentration.

The incorporation of ^{15}N was found not only in the root but also in the outer and inner leaf sheaths (Table 1). Since the ^{15}N content in the outer leaf sheath, which consisted mostly of decaying tissues, was higher than that in the root, it suggests that the outer leaf sheath does not function as a sink of nitrogenous substances translocated from roots but rather is another site of nitrogen fixation.

To check whether the activity observed in the outer leaf sheath was due to photoautotrophic or heterotrophic agents, the stem of the plant was covered with aluminum foil. The outer leaf sheath of covered plants showed higher N_2-fixing activity than that of uncovered plants (Table 1), indicating that heterotrophic nitrogen fixation took place.

Table 2 shows the distribution of $^{15}N_2$ fixed by the rice plants. A relatively high percentage of ^{15}N fixed was obtained in the root and outer leaf sheath. In Experiment A, less than 1% of nitrogen fixed was found in the leaf blade, suggesting that the fixed nitrogen is immobile in the N_2-fixing sites. In Experiment B, however, the leaf blade and young panicle received nearly 10% of the fixed nitrogen. It is not clear whether this is entirely due to translocated nitrogen.

III. CHANGES OF N_2-FIXING POPULATION AND ACTIVITY ASSOCIATED WITH WETLAND RICE

During the wet season of 1977, *in situ* acetylene-reducing activity (ARA) and N_2-fixing population associated with two rice varieties, IR26 and IR36, was examined at different growth stages.[8] Both varieties had increased activities as they grew and the highest activity was found 64 days after transplanting with 49.9 ± 4.9 and 38.7 ± 5.5 μmol C_2H_4/day per hill for IR26 and IR36, respectively. At this stage, IR26 was about 21 to 26 days before heading whereas IR36 was already at its heading stage. Activity declined 64 days after transplanting. Both varieties gave almost similar average daily ARA val-

Table 1
$^{15}N_2$ FIXATION BY RICE PLANTS GROWN IN THE FIELD THEN WATER-CULTURED IN A GAS-TIGHT CHAMBER FOR 7 DAYS

	Root	Outer leaf sheath	Basal node	Inner leaf sheath	Leaf blade	Young panicle
		Experiment A (IR26)				
Atom % excess ^{15}N	0.766	2.074	—	0.072	0.002	0.004
µg ^{15}N/plant	498	441	—	136	3.26	1.095
		Experiment B (Latisail)				
Atom % excess ^{15}N	0.501	0.897	0.245	0.031	0.033	0.020
µg ^{15}N/plant	89.3	149	20.9	30.6	26.1	8.31
		Covered				
Atom % excess ^{15}N	0.424	1.856	0.385	0.059	0.042	0.043
µg ^{15}N/plant	73.2	324	33.2	50.7	33.7	15.5

Table 2
PERCENTAGE DISTRIBUTION OF ^{15}N FIXED BY RICE PLANTS FROM ATMOSPHERE

	Root	Outer leaf sheath	Basal node	Inner leaf sheath	Leaf blade	Young panicle
Expt. A (IR26)	46.2	40.9	—	12.5	0.3	0.1
Expt B (Latisail)						
Uncovered	27.5	46.0	6.45	9.44	8.05	2.56
Covered	13.8	61.1	6.26	9.56	6.35	2.92

ues despite the differences at several growth stages. The activities at the ratoon stage were lower than those before harvest. Total ARA rates during the period of rice growth were estimated as the product of the average of daily ARA rates and days. Thus, based on a 3:1 conversion ratio, the nitrogen fixed from transplanting to harvest of IR26 (107 days) was 5.9 kg of N per hectare whereas it was 4.8 kg of N per hectare for IR36 (95 days). Under our experimental conditions, therefore, the heterotrophic N_2-fixing activity associated with rice is too low to sufficiently supply the nitrogen demand of a rice plant (40 to 60 kg N per crop). Our data also show that varietal differences in N_2-fixing activity are dependent upon the growth stage. Comparison of varieties for N_2-fixing ability at a single growth stage, hence, may be misleading.

The plate count of heterotrophs and most probable number (MPN) of N_2-fixers in or on the root seemed to increase, reaching 10^8 cells per gram fresh weight as the plant matured. No great difference was observed between IR26 and IR36 in counts of N_2-fixers in either the rhizoplane or the histosphere. Semisolid malate-yeast extract gave 10 to 100 times lower number of N_2-fixers than did semisolid glucose-yeast extract, suggesting that *Azospirillum* is less numerous in rice than in tropical grasses as reported by Döbereiner and Day.[2] Although *Azospirillum* was found also at the lower dilutions of MPN glucose-extract cultures, higher dilution cultures always confirmed the presence of Gram-negative small rods. No clear relationship between *in situ* ARA and N₂-fixing population was observed. In the analysis of this relationship, however, two factors must be reckoned with: variation of MPN and the contribution of the stem to N₂-fixation. The variance of MPN was larger than the variance of ARA rates.

IV. NEW TYPE OF N₂-FIXING BACTERIA FOUND IN WETLAND PLANTS

Using tryptic soy agar (0.1% tryptic soy broth and 1.5% Noble agar) as isolation medium, Watanabe and Barraquio[7] were able to obtain from the histosphere abundant Gram-negative small rods capable of reducing acetylene in semisolid glucose-yeast extract medium. The isolates could not grow and reduce acetylene in nitrogen-free medium. N₂-fixing activity was exhibited only when the medium was provided with a small amount of yeast extract or casamino acids or any amino acid (0.1 g/ℓ). The bacteria could grow, but could not reduce acetylene, in ammonium-supplemented (16 mg N/ℓ) medium. Initial pO₂ of 0.01 atm was optimum for nitrogenase activity. ¹⁵N₂ experiment confirmed the N₂-fixing ability of the isolates. The isolates form about 2 to 3 mm thick, white, straight pellicle 1 to 2 mm below the surface and very slight acid in semisolid glucose-casamino acids. The ARA in semisolid glucose-yeast extract (2 to 3 mℓ) is about 10 nmol C₂H₄ per tube per hour. The colonies on tryptic soy agar are white, glistening, slightly convex, entire, about 2 to 3 mm in diameter. Results of preliminary characterization and identification showed that the bacteria resemble *Achromobacter*.[3,6]

A survey of *Achromobacter*-like organisms in different wetland and dryland plants was carried out. No *Achromobacter*-like organisms were isolated from dryland plants which included *Cassia tora*, corn, sorghum, *Panicum maximum*, *Digitaria smutsii*, *Paspalum plicatum*. All the wetland plants grown on unfertilized flooded fields showed positive occurrence of this type of N₂-fixing bacteria (Table 3). The organisms were numerous in the histosphere with maximum occurrence of about 90% of the total aerobic heterotrophic N₂-fixing bacteria. Incidence of N₂-fixing bacteria in the root zone was considerably higher in wetland plants (20 to 90% of the total heterotrophic bacteria) than in dryland plants (0 to 5% of the total heterotrophic bacteria.)

V. CONTRIBUTION OF TISSUES OTHER THAN ROOT TO HETEROTROPHIC N₂-FIXATION ASSOCIATED WITH RICE

In a previous paper,[10] it was found that a portion of the N₂-fixing activity was associated with the top parts of the rice plant (Table 4). We assumed that considerable N₂-fixing activity was associated with the lower portion of the stem. In another paper,[9] the contribution of the lower part of the stem to N₂-fixation was also confirmed. The excised stem (2 to 4 cm from the tip) was again shown to exhibit N₂-fixing activity and yielded an appreciable number of N₂-fixing bacteria including *Azospirillum*.[8]

Recent experiments in water culture involving two rice varieties, IR26 and Latisail, showed that based on 5-hr ARA rates the relative contribution of the stem ranged from 100 to 6%. Heterotrophic N₂-fixation of blue green algae attached to the lower portion of the stem, and heterotrophic bacteria on the leaf sheath surface and in the nodes may explain N₂-fixing by rootless rice plants.

Table 3
POPULATION OF AEROBIC HETEROTROPHS, AEROBIC HETEROTROPHIC N_2-FIXING BACTERIA, AND *ACHROMOBACTER*-LIKE N_2-FIXING ORGANISMS IN WETLAND PLANTS[a]

Plants		Aerobic heterotrophs A	Aerobic heterotrophic N_2-fixers B (B/A)		Achromobacter-like N_2-fixing organisms C (C/B)	
				($\times 10^5$/g fresh wt)		
Aquatic weed *Monocharia vaginalis* (H), field A		10	5	(50)	4.5	(90)
Rice						
Latisail, field A	(H)	920	773	(84)	718	(93)
	(S)	300	147	(49)	66	(45)
Latisail, field B	(H)	54	11	(20)	6	(55)
	(S)	120	26	(22)	1.2	(5)
Khao-Loo, field A	(H)	2700	2025	(75)	1350	(67)
	(S)	250	148	(59)	13	(9)
IR26, field B	(H)	86	39	(45)	34	(87)
	(S)	110	22	(20)	0	
IR26, field A	(OLS)	970	146	(15)	0	
Wild rice						
Oryza australiensis (H), field A		73	55	(75)	29	(53)
	(S)	280	25	(9)	11	(44)
	(R)	240	118	(49)	106	(90)
Oryza punctata, field A	(H)	870	783	(90)	200	(26)

[a] Heterotrophic count was made on tryptic soy agar after 7 days at 30°C. One hundred colonies were randomly picked up and subjected to acetylene reduction assay in semisolid glucose-casamino acids. H = histosphere, S = stem, R = rhizome, OLS = outer leaf sheath. Field A was paddy continuously for 5 years or longer without nitrogen fertilizer and field B was converted to paddy field 1 year before.

Table 4
ACETYLENE-REDUCING ACTIVITY (μMOL C$_2$H$_4$/DAY/HILL) OF INTACT AND ROOT-CUT RICE PLANT

Growth stage	Intact	Top separated from root
	Field assay	
Trial 1 61 DAT	37.8 ± 4.5 (4)	22.7 ± 0.7 (4)
2 66 DAT	20.8 ± 3.2 (4)	13.9 ± 4.3 (4)
3 74 DAT	40.1 ± 3.7 (4)	23.0 ± 2.6 (3)
4 76 DAT	25.8 ± 3.5 (4)	24.9 ± 4.5 (3)
5 81 DAT	43.8 ± 2.8 (6)[b]	12.1 ± 1.0 (6)[b]
6 57 DAT	36.9 ± 5.6 (4)[b]	24.9 ± 1.0 (2)[b]
7 47 DAC	36.0 ± 4.5 (4)	16.8 ± 1.0 (2)
	Water culture assay	
Heading stage	22.0 ± 0.8 (4)	4.8 ± 0.5 (4)

[a]   Mean ± standard deviation of mean (replicate)
[b]   The stem portion was covered with aluminum foil to eliminate photodependent activity. DAT: days after transplanting. DAC: days after cutting.

REFERENCES

1. Balandreau, J., Millier, C., and Dommergues, Y., *Appl. Microbiol.*, 27, 662, 1974.
2. Döbereiner, J. and Day, J. M., *Proceedings of the First International Symposium on Nitrogen Fixation*, Newton, W. E. and Nyman, C. J., Eds., Washington State University, Pullman, 1976, 518.
3. Jensen, V., *Arch. Mikrobiol.*, 29, 348, 1958.
4. Lee, K. K., Alimagno, B. V., and Yoshida, T., *Plant Soil*, 42, 519, 1977.
5. Sen, M. A., *Agric. J. India*, 24, 229, 1929.
6. Ushigoshi, A., *Tsuchi to Biseibutsu*, 15, 30, 1974.
7. Watanabe, I. and Barraquio, W. L., *Nature (London)*, 277, 565, 1979.
8. Watanabe, I., Barraquio, W. L., de Guzman, M. R., and Cabrera, D. A., *Appl. Environ. Microbiol.*, 37, 813, 1979.
9. Watanabe, I. and Cabrera, D. A., *Appl. Environ. Microbiol.*, 37, 373, 1979.
10. Watanabe, I., Lee, K. K., and de Guzman, M. R., *Soil Sci. Plant Nutr. Tokyo*, 24, 465, 1978.
11. Yoshida, T. and Ancajas, R. R., *Soil Sci. Soc. Am. Proc.*, 35, 156, 1971.
12. Yoshida, T. and Ancajas, R. R., *Soil Sci. Soc. Am. Proc.*, 37, 43, 1973.

Chapter 25

USE OF *AZOLLA* AND BLUE-GREEN ALGAE IN RICE CULTIVATION IN INDIA

P. K. Singh*

TABLE OF CONTENTS

* Laboratory of Blue-Green Algae, Central Rice Research Institute, Orissa, India.

I. INTRODUCTION

Rice is the staple food of more than 60% of the population of the world and is being grown in India on about 400 million hectares where the average rice yield is around 1.32 t/ha. The introduction of high yielding varieties and other modern agricultural practices have made significant impact on rice production during recent years and there is a definite increase in the rice yield in almost all states. However, there are several constraints in obtaining higher yield and one of the important constraints is the availability of nitrogen to the rice crop in the tropical soils. The recent energy crisis has affected the availability and consumption of fertilizers, besides the economy of poor farmers is not at the stage of buying the recommended amounts. Hence, alternative sources of N are being advocated and in this respect the role of N_2-fixing photosynthetic biofertilizer is very significant. The biofertilizers like BGA are widely found in water-logged soils and there is great scope for utlizing them in increasing rice production.

Blue-green algae are widely distributed in tropical rice fields and Indian scientists[3,4,18,20] realized much earlier their importance in sustaining soil fertility. Since then there are many reports from various places in the country and abroad.[5,6,13,22,26,28] These algae contribute N more effectively when present in symbiotic association in plants like water fern *Azolla* than in the free living condition. *Azolla* has been used in N. Vietnam and China for centuries in rice cultivation and presently there is a global interest to use this aquatic plant for fertilizing rice fields.[1,10,13,15,24,25,29] The use of blue-green algae (BGA) in free living state and in symbiotic association in rice cultivation is dealt with separately in this article.

II. FREE LIVING BLUE-GREEN ALGAE

The role of BGA in the N economy of Indian rice soils is understood, although data on seasonal variation of *in situ* nitrogenase activity measured by acetylene reduction technique are not yet available. The fixed nitrogen is liberated in water/soil which is available to the rice plants. Besides providing N, these algae are also known to liberate growth promoting substances. The Central Rice Research Institute (CRRI) took the lead in India in conducting field trials on algal inoculation and crop response since 1961 at the main Institute farm and also at various locations in the country.[6,13,22] The observations of the last 18 years — the period when most of the field trials were conducted — are encouraging. Very recently a coordinated project on algae has been initiated to extend the findings of CRRI and elsewhere to the farmer's field. The major handicap is extending the use of BGA to the farmers has been due to lack of information on aspects like inoculum production, method of application, survival, extent of availability of fixed N to the rice crop, competition with native algal flora, interactions with pests, diseases, N fertilizers, pesticides, etc.

A. Inoculum Production

The production of the right type of inoculum is a very important aspect in an algal utilization program. Several methods have been worked out in various countries but most of them are sophisticated and need more investment. In the U.S.S.R, the culturing of BGA in open concrete shallow tanks has been done and a similar method is workable in Indian conditions. Shallow galvanized iron sheet trays or concrete tanks are being used in India for the production of inoculum of promising algal species. In the hot off-season, the flooded rice fields are also used for this purpose. Although a mixture of blue-green algae, *Aulosira, Anabaena, Nostoc, Gloeotrichia, Cylindrospermum, Wollea* and *Aphanothece* is suggested so that any of the algal species may grow if the conditions are suitable for its multiplication, an algal population dominated by *Aulosira* is recommended for inoculation in flooded conditions.

Shallow trays of galvanized iron sheet (4 ft × 3 ft × 9 in.) or brick and mortar structures are preferred for permanent units. The size and number of trays/tanks depends upon the amount of inoculum to be produced. It is advisable to have permanent structures with proper irrigating and draining facilities at the block level. Trays/tanks are partly filled with soil, flooded and amended with superphosphate and sodium molybdate if there is a response to molybdenum addition. Addition of lime is helpful on acidic soils in raising the pH. After the standing water becomes clear, fresh starter culture is sprinkled. The starter culture must be in healthy condition for rapid growth. Trays are kept under field condition since algae grown under such condition multiply rapidly after inoculation. The BGA multiplication during the summer is found to be more rapid than in winter and covers the area in 15 to 20 days. The grown algal mat is harvested and used for inoculation. Cultivation is continued using the amendments for 3 to 4 harvests without changing the soil. A single harvest from the tray of 1.0 m² is around 600 to 700 fresh BGA in 20 days (Tables 1 and 2). Thus it is possible to harvest about 140 kg of fresh BGA from twenty-five 8 m² trays/tanks in 20 days for inoculation of 1 to 3 ha of rice field. To prevent insect larvae which eat BGA, 3 to 5 g of diazinon/cytrolane/furaden may be added. Covering the trays/tanks with mosquito net/glass plate/transparent polythene sheet also prevents mosquito breeding. As far as possible fresh BGA should be used for inoculation purposes.[9,13,17]

Nitrogen-fixing BGA growing in fallow flooded rice fields before ploughing also can be collected and used as starter culture/field inoculant, for which some knowledge of identifying promising BGA by visual observation is required. The application of superphosphate encourages the growth of native BGA.

B. Field Application and Growth of Inoculum

The experiments conducted at CRRI have shown very clearly that fresh algal inoculation is always better than the dried ones and therefore as far as possible attempts should be made to inoculate freshly cultivated BGA under field conditions. The inoculum quantity varies in wet and dry seasons. It has been observed that algal inoculation gets damaged during heavy rains or during cloudy weather in the wet season.[9,13] Therefore, the inoculum program should be initiated after a week of planting when there is enough sunshine and no rains in clear field water, at least for a few days after inoculation. About 50 kg fresh *Aulosira* sp. (90% moisture, 3 to 5 kg dry weight) are required for inoculation during the wet season and reinoculation is recommended if there is heavy rain and growth of inoculated algae is not seen visually 7 to 15 days after inoculation. During the dry season, the establishment of inoculum is faster and therefore about 30 to 50 kg fresh weight of *Aulosira* per hectare are recommended for inoculation. The quantity of inoculum is around 135 kg fresh weight per hectare when a mixture of BGA is used. The addition of excess algal material is always better for rapid establishment. The algal material has to be applied for a subsequent 3 to 4 seasons, which might lead to establishment of inoculated algae in the field. Whether subsequent inoculation is required or not has to be checked by visual observation of appearance of BGA.

It has been observed repeatedly that application of superphosphate (20 to 40 kg P_2O_5/ha) in two split doses (during inoculation and the second one after 20 days) encouraged the growth of native and inoculated BGA (Table 3). The addition of molybdenum was not helpful in encouraging algal growth in CRRI flooded soil. Although the use of ammonium sulphate (20 to 40 kg/ha) suppresses the growth to some extent, both could be used together (as $[NH_4]_2SO_4$ in split doses) to obtain higher yield. It has

Table 1

MASS MULTIPLICATION OF NITROGEN-FIXING BLUE-GREEN ALGAE (*AULOSIRA* SP. + *A. PHANOTHECE* SP.) IN TRAYS DURING NOVEMBER — DECEMBER, 1977

Treatments				Fresh weight gm/m² tray			Dry weight gm/m² tray		
Sp.	Mo.	Lime	Pes.	*Aulosira* sp.	+ *Aphanothece* sp.	Total	*Aulosira* sp.	+ *Aphanothece* sp.	Total
1. —	—	—	—	150.0	390.0	540.0	16.16	15.42	31.58
2. +	+	—	—	185.0	420.0	605.0	20.07	16.49	36.56
3. +	+	+	—	210.0	490.0	700.0	22.51	19.18	41.69
4. +	+	+	+	165.0	380.0	545.0	17.62	15.06	32.68

Note: Soil = Clay loam   — 30 kg/m² tray;
Sp. = Superphosphate   — 4.64 gm/m² tray;
M0. = Molybdyic acid   — 0.004 gm/m² tray;
Lime   — 12.38 gm/m² tray;
Pes. = Pesticide Diazinon   — 6.19 gm/m² tray;
Inoculum — *Aulosira* sp.   — 50 gm fresh
      Aphanothece sp.   — 100 gm fresh
Incubation period   — 20 days.

Table 2
EFFECT OF VARYING DOSES OF SUPERPHOSPHATE ON MASS MULTIPLICATION OF *AULOSIRA* SP. IN TRAYS DURING APRIL — MAY, 1978

	Treatments		Fresh weight	Dry weight
	Lime. Pes.	Sp. (gm/tray)	(gm/m²)	(gm/m²)
1.	+ +	0	490	48.5
2.	+ +	12.60	670	49.0
3.	+ +	25.21	480	58.5
4.	+ +	37.81	465	45.0
5.	+ +	50.41	385	45.7
6.	+ +	63.02	375	32.6

Note: Soil — 30 kg clay loam/m² tray;
Lime — 50.41 gm/m² tray;
Pes. — Diazinon — 2.51 gm/m² tray;
Sp. — Superphosphate;
Inoculum — 30 gm fresh (3 gm dry) weight;
Incubation period — 20 days.

been observed under laboratory and field conditions that use of recommended doses of pesticides/herbicides does not affect adversely the BGA growth; on the other hand, there is rapid establishment of inoculated algae or appearance of native BGA in conjunction with pesticides since the latter kills the algal predators to some extent.[13] The growth of BGA in uninoculated fields, after a month, varied from 0.56 to 6.8 t/ha fresh weight whereas it was 1.06 to 15.6 t/ha fresh weight in 20 kg/P_2O_5 amended plots, and after BGA inoculation along with P_2O_5 amendments, it was 8.16 to 27.6 t/ha. The growth of inoculated algae after addition of 20 kg N — $(NH_4)_2SO_4$ was 7.4 to 17.4 t/ha (Table 3).

When higher doses of N fertilizers are used as a basal dose, the rice plants' growth is faster and shading occurs earlier on the water/soil surface which affects adversely the algal growth. Our recent findings show that after formation of algal mat, it is advisable to incorporate the grown BGA in soil for most benefit.[30]

C. Crop Response

Increase in crop yield is the indirect means of measuring nitrogen fixation, although all the nitrogen fixed by BGA is not available to the crop. Singh[9] estimated total N in *Aulosira* grown lawn up to 52 kg/ha but the increase in grain yield was insignificant. Our recent observations[30] showed that removal of native BGA from planted fields decreased the grain yield due to lower fertility level. There are two aspects in increasing soil fertility by growing BGA. The first one is the split application of superphosphate after planting, which encourages the growth of native BGA. In fact, in areas where N_2-fixing algae are found abundantly, it is not essential to have an algal inoculation program, and only application of superphosphate after planting is required, although BGA inoculation increases yield. The second aspect is the artificial inoculation of promising BGA alone with superphosphate in areas where the occurrence of native BGA is less. The split application of 40 kg P_2O_5/ha increased the grain yield from 6 to 28% whereas the application of 60 kg P_2O_5/ha increased the grain yield 59% over the control in our recent experiments.[30]

Initially, algal inoculation experiments started in pots. De and Sulaiman[4] did experiments for 5 years and showed that absence and presence of BGA had no appreciable

Table 3

GROWTH OF SPONTANEOUSLY APPEARING AND INOCULATED BLUE-GREEN ALGAE AND CROP RESPONSE

Treatments				Growth of spontaneous and inoculated Algae											Rice crop response Grain yield kg/ha		
				Fresh weight (t/ha)				Dry weight (kg/ha)				N content (kg/ha)					
Sp.	MO.	BGA.	AS.	1977 Wet season	1978 Dry season	1978 Wet season	1979 Dry season	1977 Wet season	1978 Dry season	1978 Wet season	1979 Dry season	1977 Wet season	1978 Dry season	1977 Wet season	1978 Dry season	1978 Wet season
1. –	–	–	–	N.D.	0.56	0.46	6.88	N.D.	12.08	50.23	97.60	N.D.	0.59	2979.00	1425.00	4173.67
2. +	–	–	–	N.D.	1.68	1.06	15.60	N.D.	60.85	161.09	213.28	N.D.	3.47	3140.25 (5.41)	1827.50 (28.25)	4585.03 (9.86)
3. +	+	–	–	N.D.	1.41	1.08	14.48	N.D.	42.26	83.80	214.00	N.D.	1.89	3024.75 (1.54)	1664.50 (16.81)	4336.58 (3.90)
4. +	+	+	–	12.20	12.60	8.16	27.60	323.36	407.16	517.46	595.36	12.29	23.32	3341.00 (12.51)	2203.00 (54.60)	5067.38 (21.41)
5. –	–	–	+	N.D.	0.36	0.20	5.60	N.D.	10.76	29.21	90.16	N.D.	0.57	3372.75 (13.22)	2661.00 (86.74)	4977.50 (19.26)
6. +	+	+	+	9.12	9.68	7.44	17.42	169.70	334.41	502.63	391.60	5.26	18.91	3561.00 (16.88)	2831.50 (98.70)	5292.19 (26.80)
Incubation period				40 d	60 d	60 d					60 d					
C.D. 5%														239.64	433.71	

Note: Plot size — k 18.6 m²; Sp. = Superphosphate (40 kg P₂O₅/ha); MO. = Molybdenum (0.6 kg Molybdic acid/ha);
AS. = Ammonium sulphate (20 kg N/ha);
BGA = Fresh inoculum of *Aulosira* sp. + *Aphanothece* sp. + *Gloeotrichia* sp. at the rate of 187 kg/ha during wet 1977 and 135 kg/ha during dry 1978, wet 1978 and dry 1979;
N.D. = Not determined;
Figures in parenthesis indicate increase over control;
Incubation period — 50 days;
Rice varieties — Dry season: Supriya;
Wet season: CR-1005;

difference on the yield up to 3 years, whereas yield was significantly higher in the fourth and fifth years. Singh[20] reported increase in yield of 368% as a result of algal inoculation whereas Sunder Rao et al.[23] reported an increase of 226% due to inoculation of *A. fertilissima*. The experiments carried out at CRRI for 3 years revealed that continuous growth of *Aulosira* increased the yield of variety IR8 210% over the control.[8]

Significant increases in the yield of paddy as a result of algal inoculation was obtained in several field trials conducted at CRRI where application of 1000 kg of lime, 100 kg of superphosphate, and 0.28 kg sodium molybdate per hectare was recommended. The increase in yield of rice was 18 and 82% after BGA inoculation alone and BGA plus fertilizer mixture (as mentioned above), respectively. Partial soil sterilization along with BGA inoculation and addition of fertilizer mixture increased grain yield by over 95% under field conditions. The increase in grain yield was overall comparable to the application of 20 kg N/ha as ammonium sulphate (Table 3). The residual effect persisted for three crops harvested in succession. The growth of the alga *Aulosira* for a month in the field before planting also increased grain yield up to 15%. Several demonstration trials were conducted in Orissa, Tamil Nadu and results were encouraging.[6,13,22]

The earlier observations on the use of BGA along with chemical N fertilizers were not encouraging. However, our recent trials showed that BGA could be used effectively along with 20 to 40 or 60 kg N when applied in split doses. Algal inoculation is also reported to be beneficial even with treatments of 100 to 150 kg N/ha and about one third of N fertilizer dose would be reduced at all N levels.[27] This was attributed to supply of growth promoting substances. However, growth of BGA is affected directly by the application of higher doses of N fertilizers[7] (Table 4) and indirectly, as boosting the growth of rice plants at an early stage reduces light for algal growth. Hence, such reports need thorough investigation. The N percentage and free amino acids are also known to increase in rice grain after BGA inoculation.

The algal inoculation program is spreading in South India and it is being used by some of the farmers. The algal material required for field inoculation to provide 20 to 25 kg N costs around U.S. $3 to 4, whereas the cost of 25 kg N is around U.S. $12.00. The cost of algal inoculation can be reduced if the trays/tanks are continuously used and BGA are also collected during the off-season from the flooded/wet fields.

III. AZOLLA

Blue-green algae fix nitrogen more effectively when present in symbiotic association in leaves of plants like *Azolla*, and therefore this water fern is considered an attractive candidate for photosynthetic production of nitrogen fertilizer. It is being used in N. Vietnam on about 800,000 ha of rice fields providing about 50% N to produce 5 ton/ha of paddy.[5] It has also been used in China for centuries and presently occupies 1,300,000 ha of subtropical and temperate rice.[1] There are six extant species of *Azolla*, namely, *A. filiculoides*, *A. caroliniana*, *A. mexicana*, *A. microthylla*, *A. pinnata* and *A. nilotica*. *A. pinnata* is commonly found in Asia and is being agriculturally used. *A. pinnata* is widely distributed in India and mostly found in shallow ponds, ditches, channels containing idle water during winter to early summer. The important aspects in the utilization of *Azolla* are the cultivation, method of application, and the response of the rice crop.

A. Cultivation

The cultivation of *Azolla* year round is important, since it multiplies mostly vegetatively, although reproductive structures are formed in plenty during winter. *A. pinnata*

Table 4

GROWTH OF INOCULATED *AULOSIRA* SP. IN RICE FIELD WITH VARYING DOSES OF N FERTILIZER DURING DRY SEASON 1979

	Treatments Ammonium sulphate + BGA (kg N/ha)	Fresh weight (kg/ha)	Dry weight (kg/ha)
1.	0	4266.7	531.8
2.	30	1653.3	158.4
3.	60	960.0	109.7
4.	90	800.0	87.5
5.	120	426.7	53.8

Note: Plot size — 20 m².
Inoculum — 50 kg/ha fresh weight.
Incubation period — 40 days.

was collected locally and is being multiplied at the CRRI farm all the year round for the last few years. It is observed that the occurrence of 5 to 10 cm standing water, and application of superphosphate at the rate of 4 to 8 kg P_2O_5/ha/harvest are essential) besides use of pesticides to protect from insect pests.[10,11,13,14,15,16,18,21] *Azolla* plants get accumulated due to wind action, and therefore fields must be divided into subplots of about 50 to 100 m² by raising the bunds. Plant sowing density of 0.1 to 0.4 kg/m² was found most favorable for rapid growth whereas thin sowing takes more time to cover the area. Soils of pH 5 to 8 supported *Azolla* growth, being optimum at neutral pH, while very acidic soils (pH 2.9 to 3.6) did not support growth. It grows well from 14 to 35°C water temperature, being optimum 20 to 30°C. Its green mass increases two to five times in a week, and in continuous multiplication trials the crop must be harvested after formation of mat to avoid decomposition. With an inoculum of 0.37 kg/m², the annual harvest was 347 t/ha/year fresh produce containing 868.5 kg N (Table 5). Concrete tanks after putting in 6 in. of soil are ideal and annually 321 t/ha of green mass was produced.[15] Application of 1 kg of P_2O_5 produced a quantity of *Azolla* equivalent to 3 kg N.[18]

Among the various isolates of *Azolla*, the *A. pinnata* isolate from Bangkok grew better than the India isolate during winter, yielding 12 t/ha, whereas the India isolate yielded 9 t/ha in 9 days using equal inocula.[18]

B. Nitrogen Contribution

The nitrogen content in Azolla is 4 to 5% on dry weight basis and 0.2 to 0.3% on fresh weight basis. The harvested *Azolla* contains about 94% water. At the CRRI farm, 30 kg N/ha/1 to 2 week is assimilated[15,18] whereas it it was 30 to 40 kg/ha/2 week at the IRRI farm.[27] About 60 kg/N/ha/45 days was reported in the U.S.[24] From continuous multiplication trials, the annual N contribution is reported as 864, 840, and 450 kg N/ha in Vietnam,[2] India,[15] and Philippines,[27] respectively.

The mineralization of *Azolla*-N is rapid and about 56% N was released as ammonia in 3 weeks of incubation at 24 + 2°C and release was faster at higher temperatures.[15] About 62 to 75% *Azolla*-N is released as ammonia in 6 weeks under IRRI conditions.[27]

The chemical composition of *Azolla* shows that it is also rich in phosphorus (0.5 to 0.9%), calcium (0.4 to 1.0%), potassium (2 to 4.5%), magnesium (0.5 to 0.65%), manganese (0.11 to 0.16%), and iron (0.06 to 0.26%) besides N, estimated on a dry weight basis,[15] and therefore also provides other essential nutrients to the rice crop.

Table 5
ANNUAL FRESH WEIGHT AND NITROGEN YIELD OF
A. PINNATA AT CRRI FARM[a]

Year	Mean water temperature (°C day-night)	Fresh weight (t/ha)			N content
		Inoculated	Harvested	Increase	
1976[b]	32.9 — 26.0	166.6	421.8	254.4	636
1977	31.3 — 24.3	160.8	527.6	366.8	917
1978	30.6 — 24.6	171.8	529.9	421.1	1052
Average	31.5 — 24.9	166.4	514.1	347.4	8685

[a] The data were collected from 32 plots with a total area of 256 m^2 and crop of Azolla was harvested generally four times in a month. Nitrogen content was calculated on 0.25% N basis.

[b] Date were collected from June to December.

C. Insect Parasites

Azolla is damaged by the larvae of insects like *Chironomus, Pyralis, Nymphulus,* etc. The control of insect pests is an important step in *Azolla* cultivation, otherwise the entire crop of *Azolla* gets decomposed rapidly. The pesticide carbofuran (65 g/ha) was found to be most effective in controlling *Azolla* pests.[14] The mixing of a lower quantity of carbofuran (3 to 30 mg/kg of *Azolla*) along with the inoculum protects the crop for a week from insect attack as well as enhances the fern growth. In fact, it is advisable to use a low quantity of pesticide regularly along with each inoculum.

D. Application in the Field

An *Azolla* multiplication program needs intensification about a month before planting, to obtain sufficient inoculum. During the off-season, particularly in summer in Northern India, its culture has to be maintained under shade, preferably in concrete tanks. *Azolla* nurseries need to be developed in villages or blocks where irrigation facilities are assured. It is worth mentioning here that some of the farmers in the villages in N. Vietnam maintain the *Azolla* culture in the off-season in small ponds enclosed by trees and sell to other farmers whenever required.

If enough water is available in the field before planting, it is grown and used as a green manure. For this purpose the bunded irrigated fields are ploughed, leveled and fresh *Azolla* plants are spread on the standing water at the rate of 500 to 1000 kg/ha along with the application of 4 to 8 kg P_2O_5/ha/week. The higher sowing density (2000 kg/ha) helps in covering the area in a short time. After 10 to 20 days of incubation it covers the area, then plants are incorporated in the soil by ploughing and rice seedlings are planted. In situations where growing *Azolla* before planting is not possible due to scarcity of water or due to high temperature, fresh *Azolla* culture (200 to 1000 kg/ha) is inoculated after establishment of seedlings along with superphosphate and pesticide, possibly in two to three split doses. It covers the area in 20 to 40 days depending upon the inoculum and then the plants are incorporated in the soil with the help of paddy weeder, after draining the field. It is not easy to incorporate *Azolla* in the presence of standing water but it gets decomposed anyway, although several times its decomposition occurred towards the maturing stage and then most of the *Azolla*-N may not be available to the same crop.

Besides growing *Azolla* before planting or along with rice plants, fallow rice fields or lands near to the rice fields also could be utilized for its cultivation, as 5 to 10% of the land can produce organic matter for fertilizing the entire area in 2 to 3 months.

Azolla and N-fertilizer could be used together as basal or as top dressing to obtain increased grain yield.

E. Effect on Rice Yield

The field experiments conducted during the last 4 years in both wet and dry season using several high yielding varieties, as well as the trials conducted at the operational research area in the villages, under the All India Coordinated Agronomic Research Project and also by the All India Coordinated Rice Improvement Project (AICRIP) at various locations, are encouraging. Both green manuring and dual cropping were used, as described below.

1. As a Green Manure

The experiments were conducted both by incorporating and using unincorporated fresh *Azolla* before planting. The unincorporated *Azolla* also gets decomposed after a certain period in the field. The height, tiller number, dry matter, number and weight of panicles, grain and straw yield, 1000 grain weight and nitrogen content at different growth periods suggested that the incorporation of one thin layer, amounting to 10 t fresh *Azolla* per hectare (94% moisture), is as efficient as the basal application of 25 to 30 kg N/ha as ammonium sulphate (Table 6). However, split application of N fertilizer was superior. An increase in grain yield was obtained up to 54% (0.5 to 1.5 t/ha) over control by incorporating 10 t/ha. A similar increase in straw yield was also observed. Varying the quantity of *Azolla* from 5 to 20 t/ha provided 12 to 50 kg N which increased the yield linearly. The response to *Azolla* green manuring was better during dry than wet seasons and short duration varieties responded better than late duration varieties. The combination of inorganic N fertilizer and *Azolla* gave better results and crop response was obtained even with the higher dose of N fertilizer, i.e., 130 kg N/ha, where *Azolla* was incorporated in the soil before planting (Table 7).[10,11,12,13,15,18,21]

2. As Dual Cropping with Rice

The dual culture of rice and *Azolla* is found to be successful.[15,18] This method appears to be easily adaptable. The fern is spread on the water surface after a week of planting along with split doses of superphosphate and pesticide. It grows along with rice plants and releases most of the fixed N after death and decay, spontaneously or after incorporation in the soil. The inoculum of 200 to 500 kg/ha established rapidly in the field and lower inoculum failed to come up. Using the 500 kg fresh weight per hectare in comparative experiments, *A. pinnata* (Bangkok) grew faster than the *A. pinnata* (India) and both covered the entire area in 20 to 25 days (Table 8). It is advisable to inoculate *Azolla* after about 10 days of planting, giving a basal dose of N fertilizer, since a thick mat of *Azolla* at an early stage might affect the tillering (Table 8). Since most of the *Azolla* N is available only after its decomposition, it is advisable to incorporate it in soil after formation of a mat. In other types of experiments, *Azolla* was cultivated separately and incorporated in soil in various quantities after a month from planting and, depending upon the quantity of *Azolla*, an increase in grain yield was obtained up to 33% over the control (Table 9). The medium and late duration varieties are more suitable for dual culture of *Azolla*. Floating *Azolla* also controls the weed growth to some extent.

F. Nitrogen Uptake and Residual Effect

The nitrogen uptake by plants was greater in the dry season crop than in the wet season and a linear increase was observed from lower (5 t/ha) to higher (20 t/ha) doses

Table 6
EFFECT OF AZOLLA[a] AND INORGANIC N FERTILIZER ON YIELD OF PADDY

	Kharif, 1976 (CR-1005)				Rabi, 1977 (Kalinga-2)			
	Grain yield		Straw yield		Grain yield		Straw yield	
Treatments	kg/ha	% increase	kg/ha	% increase	kg/ha	% increase	kg/ha	% increase
1. Control (no *Azolla*)	4875	—	4925	—	1722	—	1325	—
2. Incorporated *Azolla* (10—12 t fresh wt./ha)	5316	9	6175	25	2423	41	2087	58
3. Incorporated *Azolla* (20—24 t fresh wt./ha)	5605	15	6650	35	2623	53	2587	95
4. Unincorporated *Azolla* (10—12 t fresh wt./ha)	5158	6	5800	18	2400	39	2025	53
5. 20 kg N/ha (basal)	5173	6	5900	20	2208	28	2100	58
6. 40 kg N/ha (basal)	5483	13	6725	37	3187	85	3437	159
7. 60 kg N/ha (basal)	5819	19	6750	37	3518	104	3737	182
8. 80 kg N/ha (basal)	6082	25	6800	38	3894	126	4650	251
9. 30 kg N/ha + *Azolla* (10—12 t/ha)	5664	16	6400	30	3461	101	2837	114
10. 50 kg N/ha + *Azolla* (10—12 t/ha)	6363	31	7750	57	3576	108	3032	129
C.D. (1%)	203.58				355.78		526	

[a] 10 and 12 t/ha freshly harvested *Azolla* was used during Dry and Wet seasons respectively.

Table 7

EFFECT OF *AZOLLA* GREEN MANURING WITH HIGH DOSE OF N FERTILIZER ON THE GRAIN YIELD OF RICE CULTURE JS-52 DURING DRY SEASON, 1979

Sl.No.	Treatment	Grain yield (kg/ha)
1.	Control	1525
2.	*Azolla* (India) 20 T/ha freshly incorporated	3025 (98.4)
3.	*Azolla* (Bangkok) 20 T/ha freshly incorporated	2650 (73.8)
4.	50 kg N/ha as AS (split)	2950 (93.4)
5.	130 kg N/ha as AS (split)	5580 (265.9)
6.	130 kg N/ha as AS + *Azolla* (India) 20 t/ha incorporated	5890 (286.2)
7.	130 kg N/ha + *Azolla* (Bangkok) 20 t/ha incorporated	6080 (298.6)

Note: P application = 20 kg P_2O_5/ha in 5—7 treatments.
Plot size = 21 m².
Figures in parenthesis represent % increase over control.
AS = ammonium sulfate.

Table 8

DUAL CROPPING OF *AZOLLA* (INDIAN AND BANGKOK ISOLATES) AND RICE VARIETY KALINGA-2 DURING THE DRY SEASON, 1979

Sl. No.	Treatments	Fresh yield of *Azolla* after 25 days of inoculation (t/ha)	Plant height (cm) (53 DAT)	No. of tillers/ hill (53 DAT)	No. of tillers/ m² (53 DAT)	Grain yield (kg/ha)
1.	Control	—	43.2	5.3	240.6	2125
2.	*Azolla* (India) unincorporated	10.95	44.9	5.6	231.3	2215 (4.2)
3.	*Azolla* (India) incorporated	8.23	46.1	6.0	245.0	2515 (18.3)
4.	*Azolla* (Bangkok) unincorporated	16.33	44.4	5.6	237.6	24.90 (17.2)
5.	*Azolla* (Bangkok) incorporated	12.85	48.8	6.6	284.0	2865 (34.8)
6.	30 kg N/ha as (split)	—	45.7	7.6	316.0	2675 (25.9)
7.	30 kg N/ha as (split) + *Azolla* (India) incorporated	11.0	47.9	7.3	306.0	3365 (58.4)
8.	30 kg N/ha as (split) + *Azolla* (Bangkok) incorporated	14.66	51.8	8.0	330.3	3640 (71.3)

Note: DAT = Days after transplanting.
Plot size = 21 m².
Azolla Inoculum = 500 kg fresh Wt/ha.
P Application = 12 kg P_2O_5/ha (split) after *Azolla* inoculation.
Figures in parenthesis represent % increase over the control.

Table 9

EFFECT OF BASAL AND TOP DRESSING OF AZOLLA ON GRAIN AND STRAW YIELD

Treatments	Wet, 1977 Culture CR 191-5		Dry, 1978 Supriya		Wet, 1978 CR 191-5	
	Grain yield (kg/ha)	Straw yield (kg/ha)	Grain yield (kg/ha)	Straw yield (kg/ha)	Grain yield (kg/ha)	Straw yield (kg/ha)
1. Control	3926.7	4233.3	2097.6	1633.3	3700	3944.0
2. Azolla basal (5 tons/ha, inc.)	—	—	2714.9 (29)	1566.7	3900 (5)	3888.0
3. Azolla basal (10 tons/ha, inc.)	4736.7 (20)	4666.7 (10)	2376.0 (13)	2000.0 (22)	4200 (14)	4666.0 (18)
4. Azolla basal (5 tons/ha, uninc.)	4120.0 (5)	4666.7 (10)	2666.7 (27)	1833.3 (12)	4000 (8)	4744.0 (20)
5. Azolla basal (10 tons/ha, uninc.)	5183.3 (32)	5200.0 (23)	2768.5 (32)	2200.0 (35)	4200 (14)	4722.0 (20)
6. Azolla top dressing (5 tons/ha, inc.)	4440.0 (13)	4250.0 (0)	2505.7 (19)	1916.7 (17)	4700 (27)	5055.0 (28)
7. Azolla top dressing (10 tons/ha, inc.)	4556.7 (16)	4200.0 (0)	2643.3 (26)	1800.0 (10)	4200 (14)	4944.0 (25)
8. Azolla top dressing (15 tons/ha, inc.)	5230.0 (33)	5600.0 (32)	—	—	—	—
9. 30 kg N/ha (ammonium sulfate split)	5670.0 (44)	6000.0 (42)	2898.8 (38)	2450.0 (50)	4400 (19)	4555 (15)
C.D. 5%	975.2	N.S.	398.9	368.3	213	736

Note: Figures in parenthesis represent % increase over control.

I. INTRODUCTION

The recent discovery of grass-bacteria associations contributing to the nitrogen economy of several forage and agricultural crops prompted the attention of a number of investigators. The association of *Azotobacter* with *Paspalum* and *Spirillum* with grasses and cereal crops has now been well established.[7,8,20] The occurrence of *Azospirillum* sp. in rice soils and in association with rice roots has recently been established.[5,9,12] The remarkable ability to supply nitrogen to successive rice crops in submerged soils has been attributed to the activities of the indigenous nitrogen-fixing microorganisms.[10,14] Free-living and associative nitrogen-fixing microorganisms significantly contribute to the nitrogen economy of these soils.[16,21] For heterotrophic nitrogen fixation, the available carbon source is the major limiting factor in most of the soils. Substantial gains in N$_2$-fixation following the addition of carbon substrates have been observed.[10,17,18]

Information concerning the influence of carbon source, combined nitrogen, pesticide application and water regime on N$_2$-fixation, employing sensitive ^{15}N-tracer technique, is limited for tropical Indian rice soils. Recent studies on these lines from this laboratory, employing C$_2$H$_2$ reduction and ^{15}N$_2$ incorporation, are summarized here.

II. MATERIALS AND METHODS

A. Soil

Alluvial (pH 6.2, organic matter 1.6%, total N 0.09%, electrical conductivity 0.6 mmhos/cm) was used in the study. The soil was air-dried, screened (2 mm), and placed in glass vials (1.2 cm × 5 cm) in 5 g amounts and amended with 1% (weight per weight) powdered cellulose and thoroughly mixed with the soil prior to flooding. One set of the soil was flooded (1.5 cm standing water) and the other was held nonflooded (50% W.H.C.).

B. Effect of Ammonium Sulphate

To study the effect of fertilizer nitrogen on nitrogen fixation, 5 g portions of the soil amended with 1% cellulose were mixed with (NH$_4$)$_2$ SO$_4$ corresponding to 20, 40. 60, 80 and 100 ppm N in the vials. These were incubated under both flooded and nonflooded conditions.

C. Studies on *Azospirillum lipoferum*

Azospirillum was isolated from the soil samples incubated under submerged conditions by transferring a loopful of the soil to test tubes containing 20 mℓ sterile semisolid nitrogen-free malate medium.[8] A typical white, dense, fine pellicle developed a few millimeters below the surface of the semisolid medium within 48 hr at 30°C. Purification was achieved by serially transferring the culture to the fresh semisolid malate medium.

Azospirillum was isolated from the roots of several rice cultivars grown under field conditions in micro-plots under submerged conditions. Three plants of each variety were carefully uprooted with intact root system and the roots were thoroughly washed with tap water and the cut root pieces (0.5 to 1.0 cm) were surface sterilized with 80% ethanol. These were subsequently washed five times in sterile phosphate buffer (pH 7.0). The root pieces of uniform weight were then transferred to semisolid nitrogen-free malate medium. The organisms were identified following the culture techniques described by Neyra and Dobereiner.[13] The isolates obtained from rice soils and roots of several rice cultivars in the present study possessed the same characteristics and belonged to *Azospirillum lipoferum* (Beijerinck) Comb. nov. (Group II) based on the classification by Tarrand et al.[19]

D. Influence of Fertilizers and Pesticides on *Azospirillum lipoferum*

Technical grade benomyl (methyl-1(butyl-carbamoyl)-2-benzimadazole carbamate) and AB (2-aminobenzimadazole), a hydrolysis product of benomyl at 0, 10, 20, and 100 ppm were applied to a submerged soil. The population of *Azospirillum lipoferum* in unamended and benomyl-amended soils was estimated.[4] Nitrogenase activity was measured in these cultures employing acetylene reduction technique. The influence of NH^+_4-N and NO^-_3-N on N_2-fixation by *Azospirillum lipoferum* cultures obtained from different rice cultivars with clearly demarcated low, medium, and high N_2-ase activity was studied employing C_2H_2 reduction assay.

E. ^{15}N-Studies

The soil samples amended with 1% cellulose and the five concentrations of ammonium sulphate were transferred to a desiccator wrapped with black paper to prevent algal growth. Excess CO_2 produced during incubation was absorbed by 40% KOH placed in the desiccator. Wet cotton moistened with distilled water was kept in a few separate vials for maintaining humidity in the desiccator. This was then sealed and the air was evacuated by vacuum and flushed three times with argon. The desiccator was then filled with a gas phase which consisted of 0.2 at O_2, 0.5 at Ar and 0.3 at N_2 with 72.5 at% excess ^{15}N (Prochem, London). The system was then incubated for 28 days at 28°C in the dark. At the end of incubation, the total N in the soil samples was determined by the Kjeldahl method and the distillates were used for the determination of ^{15}N enrichment in the soil samples.

Several purified *A. lipoferum* cultures obtained from different rice cultivars were inoculated individually in 30 mℓ of semisolid N-free malate medium contained in special 100-mℓ vacuum flasks. These were then exposed for 4 days to an atmosphere containing oxygen (0.2 at), argon (0.5 at) and nitrogen (0.3 at) with 72.5 atom % excess ^{15}N$_2$. The isotopic ratio analysis was carried out at the Seibersdorf Laboratory of the International Atomic Energy Agency, Vienna.

III. RESULTS AND DISCUSSION

The isotopic analysis of the soil samples indicated significant indigenous nitrogen-fixing activity even in the absence of cellulose. Soil submergence favored nitrogen fixation. Results suggest that application of cellulose stimulated nitrogen fixation under both flooded and nonflooded conditions irrespective of soil properties. It has been established that under flooded conditions, products of anaerobic decomposition may serve as energy sources for aerobic nitrogen fixers,[11] thus favoring the activities of both aerobic and facultative anaerobic nitrogen fixers in such conditions.

Combined nitrogen is known to inhibit nitrogen fixation in pure culture and in flooded soil ecosystems.[3,15] Our studies with ^{15}N$_2$ indicate a gradual decrease in nitrogen fixation with increase in the concentration of NH_4^+-N in the alluvial soil under both flooded and nonflooded conditions (Table 1). Moreover, the inhibition is more pronounced beyond 60 ppm-N under nonflooded conditions. Contrary to this, in the acid sulphate (pokkali) and the acid saline (karapadam) soils nitrogen fixation was completely suppressed in the presence of ammonium sulphate under both flooded and nonflooded conditions.[6] These results suggest that the degree of inhibition of nitrogen fixation by ammonium sulphate depends on the soil properties. Nitrogenase activity was not completely suppressed even at concentrations as high as 160 to 320 ppm in

Table 1
NITROGEN FIXATION IN
CELLULOSE-AMENDED
ALLUVIAL SOIL AS INFLUENCED
BY WATER TREATMENTS AND
AMMONIUM SULPHATE

N-ppm	Atom % excess ^{15}N at the end of the incubation[a]	(SD)	Nitrogen fixed (mg/kg)
	Flooded		
0	0.753	0.045	12.8
20	0.648	0.017	11.0
40	0.491	0.132	8.3
60	0.423	0.069	7.2
80	0.398	0.138	6.8
100	0.383	0.083	6.5
	Nonflooded		
0	0.819	0.149	13.9
20	0.800	0.051	13.6
40	0.655	0.056	11.0
60	0.502	0.066	8.5
80	0.110	0.027	2.0
100	0.058	0.022	1.0

[a] LSD between fertilizer levels at P 1 and 5% are 0.17 and 0.12, respectively.

wet-land rice soil[15] while Yoshida et al.[22] observed complete inhibition of nitrogen fixation when the fertilizer nitrogen was applied at 160 ppm-N. The differences with respect to the inhibitory nature of the ammonium sulphate on nitrogen fixation could be attributed to soil properties and moisture regime.

We investigated the $^{15}N_2$ incorporation in *A. lipoferum* strains obtained from the roots of several rice cultivars. Albrecht et al.[1] observed that nitrogenase activity and pellicle formation were always greater in the root samples than in soil samples. Becking,[2] employing $^{15}N_2$, observed appreciable incorporation in the nitrogen-free enrichment cultures. Using 43 atom % ^{15}N the cells contained 0.43 atom % ^{15}N and with 65 atom % $^{15}N_2$ the labeling of the cells increased to 1.04 atom % ^{15}N. The $^{15}N_2$ incorporation by *A. lipoferum* cultures isolated from different rice cultivars, grown under uniform field conditions, in the present study using 72.5 atom % $^{15}N_2$, ranged from 0.205 to 11.8 atom % ^{15}N (Table 2). Evidently, strains from different rice cultivars varied markedly in their ability to incorporate $^{15}N_2$. All the isolates seemed to belong to *A. lipoferum* as per their morphological and certain physiological characteristics.

Nitrogen fixation in cultures of *Azospirillum* sp. isolated from twenty rice cultivars varied uniformly irrespective of the growth stage of the plant.[23] Since *A. lipoferum* is very closely associated on or inside the root tissue, the differential nitrogen fixation could be due to qualitative and quantitative differences in the root environment from which these were obtained. It is also known that the amount of root exudates and available nutrients greatly differ among the rice cultivars. Since there existed marked differences between the N₂-fixing capacity of different strains this opens up some possibilities for the selection of the most efficient strains and rice cultivars.

Table 2
[15]N_2 INCORPORATION BY
AZOSPIRILLUM LIPOFERUM FROM
DIFFERENT RICE CULTIVARS

Rice cultivar from which *A. lipoferum* was isolated	Nitrogen fixed mg/30 ml medium[a]	Atom % excess ^{15}N at the end of incubation[b]
Kalinga-1	15.7	11.84 ± 0.90
IR-8	13.0	9.77 ± 1.02
MR 1550	6.50	4.89 ± 0.83
Pankaj	4.80	3.60 ± 0.62
Jayanthi	2.80	2.10 ± 0.51
Pusa	2.40	1.83 ± 0.02
Padma	0.30	0.21 ± 0.05

[a] Calculated from total N determined by Kjeldahl assay.
[b] Values are means ± SE.

Table 3
INFLUENCE OF
AMMONIUM ON C_2H_2
REDUCTION BY
AZOSPIRILLUM
LIPOFERUM STRAINS
WITH VARYING N_2-ASE
ACTIVITY FROM RICE
ROOTS

	C_2H_4 n mol/ml/hr		
N-ppm	Low[a]	Medium[a]	High[a]
0	192	1312	1782
5	250	1281	1968
10	265	1781	2250
20	162	1000	1312
50	30	156	250

[a] Groups classified according to the nitrogenase activity from different rice cultivars.

Several isolates were tested for their N_2-ase activity and *A. lipoferum* strains were grouped into three (low, medium, and high N_2-ase activity) depending upon their N_2-ase activity. The influence of NH_4^+-N and NO_3^--N was tested on the N_2-ase activity of these strains. Results indicate that low levels of NH_4^+-N stimulated N_2-ase in all the cultures, while at higher concentrations (over 10 ppm) significant inhibition of N_2-ase activity occurred (Table 3). Interestingly, NO_3^--N stimulated N_2-ase activity in low-fixing group even at concentrations of 50 ppm, whereas in the other two groups the N_2-ase activity is not drastically effected (Table 4).

Striking stimulation of the *Azospirillum* population was noticed in benomyl-amended submerged paddy soils.[4] N_2-fixation was high in cultures obtained from benomyl-amended soils. Further studies with AB (a hydrolysis product of benomyl) in-

Table 4
EFFECT OF NITRATE ON C$_2$H$_2$ REDUCTION BY *AZOSPIRILLUM LIPOFERUM* STRAINS WITH VARYING N$_2$-ASE

C$_2$H$_4$ n mol/ml/hr

N-ppm	Low[a]	Medium[a]	High[a]
0	192	1312	1782
5	287	1031	1675
10	308	1156	1656
20	382	1437	1718
50	220	1281	1718

[a] Groups classified according to the nitrogenase activity, from different rice cultivars.

Table 5
POPULATION AND N$_2$-ASE ACTIVITY OF *AZOSPIRILLUM LIPOFERUM* ISOLATED FROM AB-AMENDED SUBMERGED SOIL

	Soil incubation (days)			
	10		30	
Concentration (AB) ppm	Population[a]	N$_2$-ase of *A. lipoferum*[b]	Population[a]	N$_2$-ase of *A. lipoferum*[b]
0	1.2	187	3.5	767
10	2.2	22	20	238
20	3.2	9.1	16	685
100	5.4	23.4	9.2	48

[a] Population expressed as MPN × 10^6/g dry soil. Means of five replicates.
[b] N$_2$-ase activity of isoalted *A. lipoferum* expressed as n mol of C$_2$H$_4$/ml/day.

dicated that this compound also stimulated the *Azospirillum* population even at a concentration of 100 ppm, up to 30-day incubation. Although AB stimulated *A. lipoferum* population, a progressive fall in the stimulation was noticed with increasing concentration at 30-day sampling (Table 5). Interestingly, the N$_2$-ase activity was low in *A. lipoferum* cultures isolated from AB amended soils. These results suggest that although the proliferation of *A. lipoferum* is favored by AB amendment the N$_2$-ase activity of the isolated cultures is inhibited.

ACKNOWLEDGMENTS

The authors thank Dr. H. K. Pande, for encouragement and facilities. This study was supported by the International Atomic Energy Agency, Vienna, through a research contract.

REFERENCES

1. Albrecht, S. L., Okon, Y. and Burris, R. H., *Plant Physiol.*, 60, 528, 1977.
2. Becking, J. H., *Antonie van Leeuwenhoek, J. Microbiol Serol.*, 29, 326, 1963.
3. Becking, J. H., International Atomic Energy Agency, Vienna, 1971.
4. Charyulu, P. B. B. N. and Rao, V. R., *Curr. Sci.*, 47, 822, 1978.
5. Charyulu, P. B. B. N., Ramakrishna, C., and Rao, V. R., 1981, in press.
6. Charyulu, P. B. B. N. and Rao, V. R., *Soil Sci.*, 1979, in press.
7. Day, J. M., Neves, M. C. P., and Döbereiner, J., *Soil Biol. Biochem.*, 7, 107, 1975.
8. Döbereiner, J., Marriel, I. E., and Nery, M., *Can. J. Microbiol.*, 22, 1464, 1976.
9. Lakshmikumari, M., Kavimandan, S. K., and Subba Rao, N. S., *Indian J. Exp. Biol.*, 14, 638, 1976.
10. MacRae, I. C. and Castro, T. F., *Soil Sci.*, 103, 277, 1967.
11. Magdoff, F. R. and Bouldin, D. R., *Plant Soil*, 33, 49, 1979.
12. Nayak, D. N. and Rao, V. R., *Arch. Microbiol.*, 115, 359, 1977.
13. Neyra, C. A. and Döbereiner, J., *Adv. Agron*, 29, 1, 1977.
14. Rao, V. R., Kalininskaya, T. A., and Miller, U. M., *Mikrobiologiia*, 42, 729, 1973.
15. Rao, V. R., *Soil Biol. Biochem.*, 8, 445, 1976.
16. Rao, V. R., *Curr. Sci.*, 46, 118, 1977.
17. Rao, V. R., *Soil Biol. Biochem.*, 10, 319, 1978.
18. Rice, W. A., Paul, E. A., and Wetter, L. R., *Can J. Microbiol.*, 13, 829, 1967.
19. Tarrand, J. F., Krieg, N. R., and Dobereiner, J., *Can. J. Microbiol.*, 24, 967, 1978.
20. von Bulow, J. W. F. and Dobereiner, J., *Proc. Nat. Acad. Sci. U.S.A.*, 72, 2383, 1975.
21. Watanabe, I., Lee, K. K., Alimagno, B. V., Sato, M., Del Rosario, D. C., and de Guzman, M. R., International Rice Research Institute Research Paper, ser. No. 3, Los Banos, Laguna, Philippines, 1977.
22. Yoshida, T., Roncal, R. A., and Bautista, E. M., *Soil Sci. Plant Nutr. Tokyo*, 19, 117, 1973.
23. Nayak, D. N. and Rao, V. R., unpublished data.

Chapter 27

MINERAL NUTRITION AND N_2-FIXATION IN *Azolla**

E. Malavolta,** ***W. R. Acorsi,** A. P. Ruschel*** F. J. Krug,***
L. I. Nakayama,** and I. Eimori,**

TABLE OF CONTENTS

*   The support of CNEN and FAPESP is acknowledged.
** Escola Superior de Agricultura, Luiz de Queiroz, Piracaba, S. P., Brazil.
***Centro de Energia Nuclear na Agricultura, University of Sao Paulo, Piracicaba, S.P., Brazil.

I. INTRODUCTION

Azolla spp. are well known small aquatic ferns of wide distribution being more abundant, however, in the tropics. Blue green N_2-fixing algae, usually belonging to the genera *Nostoc* or *Anabaena* are found in a chamber located in the upper lobe of the leaflets. The role of the association in improving the fertility of rice soils has been reviewed recently.[2] Rates of N_2-fixation estimated by the acetylene reduction method range from 100 to 600 kg/ha/year. Mostly the fixed nitrogen is not released until the death and decay of the organisms.

A search in the available literature has produced only one paper dealing with the influence of mineral nutrition on N_2-fixation. Watanabe et al.,[6] using nutrient solutions, showed that deficiencies of P, K, Ca, and Mg adversely affected dry matter production and nitrogen accumulation. The present contribution had the objectives of studying the effects of mineral deficiencies and excess on growth, N_2-fixation, and chemical composition of *Azolla filiculoides* under controlled conditions.

II. MATERIALS AND METHODS

Azolla filiculoides was obtained from a fish pond in Jaboticabal, S.P., Brazil, courtesy of Agricultural Engineer R. Pitelli. Of inocculum (fresh material), 5 mℓ were placed in shallow (25 × 30 × 5 cm) plastic trays containing the treatments given in Table 1. The solutions were continuously aerated and N was provided only by air.

Three weeks after the beginning of the treatments the following determinations were made: nitrogenase activity (acetylene reduction), dry matter, and chemical analysis. Symptoms shown by the organism were also described.

III. RESULTS AND DISCUSSION

A. Symptomatology

Complete nutrients — Prevailing color is moss-green. Color intensity is somehow related to the number of *Anabaena azollae* contained in the leaflet chambers. Roots, cylinder-shaped and brownish in color.

Minus P — Chlorosis in the older leaflets progressing to the newer ones. Later the leaflets became light brown and crinkled and detached easily. Roots thin, dark-brown and easily detached.

Minus K — Plants much smaller than in the "complete" treatment, branches and leaflets are reduced in size. According to age the leaflets are light green to whitish brown. *Anabaen* containing chambers reduced in size. Dark brown roots.

Minus Ca — Growth relatively normal. Younger leaflets light green and then brownish. Roots thinner, lighter in color and shorter than in the "complete" treatment.

Minus Mg — Leaflets light green and then brown. Area corresponding to chamber, blue-green. Most of the roots loose in the solution, brownish, thin and short.

Minus S — Leaflets in the tip of the branches moss-green in color; more distant ones yellowish to yellow-orange. Roots light brown.

Minus Fe — Leaflets slightly light green. Roots thinner, normal in length, whitish.

Excess Al — Individuals much smaller; crinkled leaflets, brown in color; complete absence of green pigments. Roots shorter and light brown. Strands of *Anabaena* with discoloration and separation of individual cells.

Excess Mn — Leaflets initially chlorotic, then whitish and brown. Roots brown, and relatively thicker, easily detached from the plant.

Table 1

COMPOSITION OF THE NUTRIENT SOLUTIONS (ml/l)[a]

Stock solutions	Complete	−P	−K	−Ca	−Mg	−S	−Fe	−Mo	+ Al	+ Mn
K$_2$SO$_4$ 0.5 M	5	5	—	5	5	—	5	5	5	5
MgSO$_4$ M	2	2	2	2	—	—	2	2	2	2
Ca(H$_2$PO$_4$)$_2$ 0.5 M	10	—	10	—	10	10	10	10	10	10
CaSO$_4$ 0.01 M	200	200	200	—	200	—	200	200	200	200
Fe — EDTA	1	1	1	1	1	1	—	1	1	1
Solu. a.	1	1	1	1	1	1	1	—	1	1
Na$_2$SO$_4$ 0.5 M	—	—	5	—	4	—	—	—	—	—
KH$_2$PO$_4$ M	—	—	—	1	—	—	—	—	—	—
MgCl$_2$	—	—	—	—	—	—	—	—	—	—
Solu. a — Mo	—	—	—	—	—	—	—	1	—	—
Solu. Al	—	—	—	—	—	—	—	—	2.5	—
Solu. Mn	—	—	—	—	—	—	—	—	—	50

[a] + Al, and + Mn = excess element, stock solution Al = 24 ppm; stock solution Mn = 1.86 g MnCl$_2$ 4H$_2$O/l

Table 2

EFFECT OF MINERAL NUTRITION ON DRY MATTER PRODUCTION BY *AZOLLA* (AVERAGE OF TWO REPLICATES)

Treatment	Dry matter g/tray
Complete	1.998
−S	2.493
−Mo	1.847
−Fe	1.279
+ Mn[a]	1.039
−Ca	0.964
−P	0.631
−K	0.405
−Mg	0.169
+ Al	
(0.024 ppm)	0.405
(0.060 ppm)	0.074
(0.120 ppm)	0.054

[a] Excess Mn.

B. Dry Matter

Table 2 gives the effects of treatments on dry matter production.

All "minus" treatments, with the exceptions of −S, and −Mo reduced yield. These effects could be traced to some extent to the levels of the elements found in the organisms. Mo concentration in the dry matter in the −Mo treatment was practically the same found in the control plants. Since no attempt was made to get rid of contaminants in distilled water, salt, containers, and air, impurities therein probably supplied enough of this micronutrient for the growth of *Azolla*. The effect of omitting S from the

Table 3

MINERAL CONTENT OF *AZOLLA* (AVERAGE OF TWO REPLICATIONS)[a]

Treatment	%						(ppm)							
	N	P	K	Ca	Mg	S	B	Cu	Fe	Mn	Mo	Zn	Al	Na
Complete	5.55	1.23	6.54	0.33	0.24	1.73	25	17	843	23	9	56	248	93
−P	3.55	0.10	3.41	0.27	0.22	0.97	—	22	2029	23	—	99	163	165
−K	4.23	2.04	0.07	0.63	0.55	1.47	—	19	100	38	—	101	268	1220
−Ca	3.65	1.28	5.32	0.25	0.30	1.87	—	17	654	27	—	88	182	1400
−Mg	4.30	1.23	2.55	1.33	0.10	0.85	—	22	3038	89	—	102	362	657
−S	5.30	1.12	7.07	0.35	0.29	0.47	—	17	797	27	—	49	<25	120
−Mo	5.50	1.11	5.03	0.37	0.20	1.42	—	22	221	27	7	81	179	93
+Mn[b]	4.83	1.20	7.02	0.48	0.26	1.05	—	19	700	314	—	66	217	151

[a]   Excess Al: insufficient material for analysis.
[b]   Excess Mn.

medium resulted in higher yield, a puzzling result; in fact, as shown in next section, the only relevent differences found in the minus sulfur plants when compared with those in the "complete" treatment is the Al level. *Azolla*, as Table 2 demonstrates, seems to be highly susceptible to aluminum toxicity. Al concentration in the −S individuals, which give the highest dry matter production, was the lowest found in the experiment.

With respect to the effects due to the absence of P, K, Ca, and Mg, the results are in agreement with the findings of Watanabe et al.[6]

C. Mineral Composition

The influence of the treatments on the mineral composition of *Azolla* is given in Table 3. As a rule, when a given element was omitted or in excessive concentration, there was a variation in its content in the same direction. Other points should be underlined.

1.   N content was lowest in treatments −P, −K, −C, −Mg, and +Mn wherein nitrogenase activity was also reduced (see next)
2.   The same type of general interactions found in higher plants seem to occur in *Azolla* in so far as uptake and concentration in the dry matter are concerned
3.   A significant, positive correlation was found to exist between dry matter production and N content, as shown in Figure 1

D. Nitrogenase Activity

Statistical analysis of the data in Table 4 revealed the effect of mineral nutrition on nitrogenase (Nase) activity as being significant at the 1% level (Table 5).

The treatments which significantly reduced N-ase activity were, in decreasing order, the following: −S, excess Mn, −P, −Ca, −K, and −Mg. Therefore the puzzle with regards to sulfur continues. Acid labile sulfur is part of the enzyme systems; its deficiency induced less activity not being able, however, to depress growth, which seems to indicate that S levels in the tissue required for maximum N$_2$-fixation are much higher than the needs for dry matter production. The effect of excess Mn could probably be explained by assuming a displacement of iron from the enzyme system: this has been found to be the case in higher plants with other Fe requiring enzymes.[3] The needs of both P and Mg for the phosphorylation reactions which provide energy for the endergonic reaction of N$_2$ activation explain the reduction in activity found in the corre-

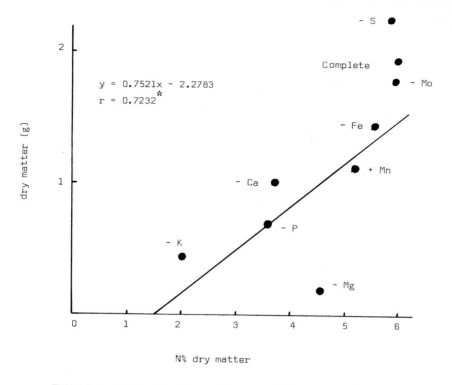

FIGURE 1.   Relationship between N content and dry matter production.

Table 4
NITROGENASE
ACTIVITY IN
***AZOLLA* AS**
AFFECTED BY
MINERAL
NUTRITION

(μg N/g/dry matter/
hr; average of 4
replicates)

Treatment	Activity
Complete	389
−Mo	382
msFe	335
−S	194
+ Mn[a]	179
−P	58
−Ca	57
−K	52
−Mg	21
[a] L . S . D . (5%).	181

Table 5
ANALYSIS OF VARIANCE ON THE
EFFECT OF MINERAL NUTRITION IN
NITROGENASE ACTIVITY IN *AZOLLA*

Source of variation	Degrees of freedom	Sum of square	Mean square	F
Treatment	8	717,426	89,678	14, 5[a]
Residue	27	156,642	5,801	
Total	35	877,068		

[a] Significant 1% level.

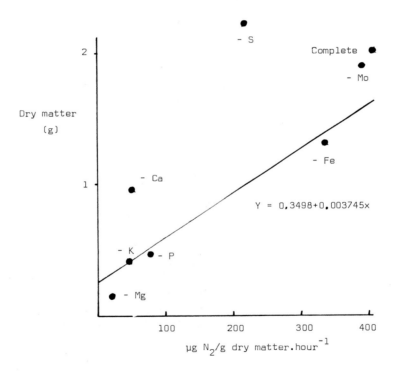

FIGURE 2.   Relationship between N-ase activity and growth of *Azolla*.

sponding treatments. Calcium, besides being necessary for heterocysts[5] is also required for membrane formation and functioning and, as a consequence, is involved in trapping light energy in chloroplasts, this being a requirement for N$_2$-fixation.[1] Potassium is involved in many metabolic processes, including phosphorylation reactions and protein synthesis[4]: its role on N-ase activity is likely to be an indirect one.

A direct relationship between N-ase activity and growth is indicated in Figure 2.

REFERENCES

1. **Burris, R. H.,** Methodology, in *The Biology of Nitrogen Fixation,* Quispel, A., Ed., North Holland, Amsterdam, 1974.
2. **Day, J. M. and Witty, J. F.,** Novel aspects of nitrogen fixation, *Outlook Agric.,* 9(4), 180, 1977.
3. **Hewitt, E. J. and Smith, T. A.,** *Plant Mineral Nutrition,* English Universities Press, London, 1975.
4. **Malavolta, E.,** *Manual de Química Agrícola — Nutrição de Plantas e Fertilidade do Solo,* Editora Agronômica Ceres, São Paulo, 1976.
5. **Mulder, E. G. and Brotonegro, S.,** Free living heterotrophic bacteria, in *The Biology of Nitrogen Fixation,* Quispel, A., Ed., North Holland, Amsterdam, 1974.
6. **Watanabe, I., Espinas, C. R., Erja, N. S. B., and Alimasno, V. B.,** The Utilization of *Azolla-Anabaena* Complex as a Nitrogen Fertilizer for Rice, International Rice Research Institute, Los Baños, Laguna, Philippines, 1977.

Chapter 28

NITROGEN FIXATION BY *AZOLLA-ANABAENA* IN CULTURE SOLUTION

Marli de F. Fiori* and A. P. Ruschel**

TABLE OF CONTENTS

* Graduate student: acknowledgement is made of a FAPESP fellowship.
** Centro de Energia Nuclear na Agricultura, Piracicaba, S.P., Brasil.

I. INTRODUCTION

The association *Azolla-Anabaena* (A-A) was recorded for the first time by Strasburger in 1873, who observed that *Anabaena azollae* was present in the dorsal lobes of *Azolla filliculoides.* The demonstration that this system was able to grow in media without mineral-N provided evidence of the N_2-fixation process.[1]

The importance of this association is that it is characterized by the high efficiency of the system,[4,5,6,7] which makes it possible to use A-A as a common practice for N-fertilization in flooded regions, in addition to which it can be used as food for chickens.[10]

The development of A-A depends upon environmental factors such as pH and combined-N. There is some controversy as to whether nitrogenase activity (ARA) is inhibited by mineral-N. Stewart et al.[9] presumed the inhibition of nitrogenase by combined-N; however Bone[2] found only partial inhibition or decreasing ARA.

As Ashton and Walmsley[1] observed, the effect of A-A in natural conditions decreases the level of water aerobiosis by reducing O_2 mixing. This work had the objective of studying the effect of solution aeration, different sources of N, and pH, together with the effectiveness of the A-A association as a nitrogen source for irrigated rice.

II. MATERIAL AND METHODS

Experiments (1 to 4) were set up in the greenhouse. In the first three experiments nutrient solution was used, and in the fourth the effect of A-A on the N-economy of rice was studied. Ferns for Experiments 1, 2, and 3 were grown in plastic pots with 1.5 l of modified Hoagland nutrient solution, being initially "sown" with ten ferns. In Experiment 1, the effect of aeration, NO_3-N and growth of the association were studied in river-water with the following mineral composition: (NH_4^+ — 0.44; NO_3^- — 1.09; Cl^- — 29.75; SO_4^- — 17.25; PO_4^- — 1.40; Ca — 4.80; Mg — 1.70; Fe — 1.00; Cu — 0.01; Mn — 0.01; Zn — 0.30; Na — 30.0; K — 6.40, and NO_2^- — 0.058). The treatments were (1) 0.3 mM of NO_3^-; (2) without N, and (3) river-water, each treatment being replicated three times with and without aeration.

In Experiment 2, the pH (4.0 — 5.0 — 5.5 — 6.0 — 6.5 and 7.0) effect was studied. Nutrient solution ($MgSO_4 \cdot /7H_2O$ — 0.492 g; $CaCl_2$ — 0.275 g; FeEDTA — 1 ml; micronutrient — 1 ml; H_2O — 976 ml) was buffered with classical (KH_2PO_4/Na_2HPO_4) phosphate buffer, and the final pH obtained by adding NaOH (0.1 N) or H_2SO_4 (0.05 N).

In Experiment 3, the effect of N-sources (NH_4^+, NO_3^- and N-urea) were studied by adding 0.7 µg N/ml as $(NH_4)_2SO_4$, NaNO_3, $(NH_3)_2CO$, and $(NH_4)_2SO_4$ + NaNO_3 1:1 N; pH — 6.0.

Ferns from Experiments 1, 2, and 3 were harvested 15 days after the start of the experiment, when ARA was evaluated by acetylene reduction (the ferns were put into a vessel and 10% of the atmosphere was replaced by acetylene). Weight (mg) and total-N were analyzed.

In Experiment 4 with A-A and rice (var. IAC-435), the rice was sown in vermiculite moistened with 10^{-4} M $CaSO_4$ solution. After 15 days the plants were transplanted to pots with soil fertilizer with P and K (100 and 50 kg/ha of P_2O_5 and KCl, respectively). The following treatments were used: (1) without N-added, (2) N-deficient (20 kg N/ha), (3) treatment (1) plus *Azolla,* (4) treatment (2) + s *Azolla,* and (5) *Azolla* incorporated in the soil. Nitrogen (206 mg NaNO_3/pot) was used and the same amount of N was added as *Azolla* for treatment (5), calculated from the N in the fern (12.4 g of green fern/pot). The rice was irrigated for 30 days, when the water was not replaced

Table 1
EFFECT OF AERATION, NO_3^- ADDITION TO CULTURE SOLUTION AND RIVER-WATER ON THE WEIGHT, TOTAL-N, AND NITROGENASE ACTIVITY IN *AZOLLA-ANABAENA* ASSOCIATION

Treatment	Weight (g)	Total-N (mg)	N-ase activity C_2H_4 (μmol/pot/hr)
Deficient-N	0.724 a	40.770 a	4.114 a
+ N	0.673 a	40.201 a	4.015 a
River-water	0.160 b	4.236 b	0.139 b
LSD (Tukey 5%)	0.121	6.454	1.809
− aeration	0.553 a	30.629 a	2.822 a
+ aeration	0.486 a	26.176 b	2.691 a
LSD (Tukey 5%)	0.081	4.335	1.215
C.V. %	12.250	17.804	51.429

Note: Means with different letters, in the same line or column of averages, are significantly different at 5% level of significance by Tukey test.

in order to permit decomposition and N-addition of *Azolla* from treatments (3) and (4). After 25 days plants were harvested and analyzed for weight, and N, using the Kjeldahl method.

III. RESULTS AND DISCUSSION

Table 1 shows the effect of aeration on the (*A-A*) association grown in nutrient solution and also the effect of river-water. Although aeration had no influence on the weight and nitrogenase activity of the *A-A* association, it was noted that nonaerated plants showed a higher total-N than the aerated. Comparison of plants grown in N-free solution or with 0.3 mM NO^{-3}, showed that there was no change in weight, total-N or nitrogenase acitivity.

However, plants grown in river-water gave only one quarter of the yield and one tenth of the total-N content compared with plants grown in nutrient solution either with or without N. As the N-level in the river-water was low, this effect might be due to excess of Cl, SO_4^- and Na, which according to Ashton and Walmsley[1] would not only permit the growth of the more tolerant plants. The conclusion was that this river-water cannot be used for *A-A* culture under normal conditions. Renault et al.[8] noted that *Azolla* does not survive in polluted rivers.

The effect of pH on *A-A* symbiosis was studied using six levels (4.0, 5.0, 5.5, 6.0, 6.5, and 7.0) and is shown in Table 2. It was observed that although the weight of the plants decreased with high pH (7.0), nitrogenase activity was higher with pH 6.0 and 6.5, decreasing with the other levels studied. Total-N, however, was higher with pH 5.5, which is in agreement with Watanabe et al.[11] who suggests this pH as optimum for normal plant development. The difference was not statistically significant.

Comparing different sources of NO_3^- and NH_4^+, it was noted that only NO_3^- decreased the weight, total-N and ARA of the *A-A* association (Table 3). This might be due to an indirect effect of Ph, as it was around pH 6.5 with $NaNO_3$ and pH 5.5 with the other treatments.

Table 2
EFFECT OF NUTRIENT SOLUTION pH ON WEIGHT, TOTAL-N, AND NITROGENASE ACTIVITY OF THE *AZOLLA-ANABAENA* ASSOCIATION

pH	Weight (g)	Total-N (mg)	N-ase activity C_2H_4 (μmol/pot/hr)
4.0	0.767 a	30.752 a	10.889 b
5.0	0.758 a	30.623 a	8.652 bc
5.5	0.790 a	36.887 a	7.874 c
6.0	0.738 a	31.308 a	17.429 a
6.5	0.679 ab	31.493 a	13.311 ab
7.0	0.558	31.392 a	11.828 b
LSD (Tukey 5%)	0.142	11.750	4.672
C.V. %	8.845	16.672	17.900

Note: Means with different letters, in the same line or column of averages, are significantly different at 5% level of significance by Tukey test.

Table 3
EFFECT OF DIFFERENT SOURCES OF N ON THE WEIGHT, TOTAL-N AND NITROGENASE ACTIVITY OF THE *AZOLLA-ANABAENA* ASSOCIATION IN NUTRIENT SOLUTION

Treatment	Weight (g)	Total-N (mg)	N-ase activity (log × + 1) C_2H_4 (μmol/pot/hr)
−N	0.507 a	18.822 a	1.214 a
NH_4NO_3	0.474 a	16.565 a	1.206 a
$NaNO_3$	0.316 b	6.042 b	0.041 b
$(NH_4)_2SO_4$	0.348 ab	11.642 ab	0.895 a
$CO(NH_2)_2$	0.401 a	13.890 a	1.066 a
$NaNO_3$ + $(NH_4)_2SO_4$	0.501 a	16.990 a	1.095 a
LSD (Tukey 5%)	0.174	8.738	0.395
C.V. %	18.269	27.723	19.159

Note: Means with different letters, in the same line or column of averages, are significantly different at 5% level of significance by Tukey test.

According to Peters and Mayne,[5,6] nitrate and urea decrease nitrogenase activity in *Azolla* by as much as 30% in relation to the reduction rate of *Azolla* grown in N_2. The results shown in Table 3 indicate that NO_3^- drastically decreased ARA, followed by ammonium sulphate and urea with much lesser rates (73 and 83%) in relation to plants grown in N-free solution.

Table 4 shows the effect of *A-A* on rice production. It is clear that the greatest yield is observed when *Azolla* was incorporated into the soil, indicating the efficiency of *Azolla* fertilization in rice crops. The variables analyzed in the N-deficient plus *Azolla* treatment were low although not statistically significant in relation to the others, except for treatment with incorporation of *Azolla*, which suggests nutrient absorption competition between *Azolla* and rice. ARA of rice roots was nil in all treatments.

Table 4
EFFECT OF *AZOLLA-ANABAENA* ON THE WEIGHT, AND TOTAL-N OF RICE GROWN IN WATERLOGGED SOIL (LVa)

Treatment	Weight (g/pot)		Total-N (mg/pot)	
	Shoot	Root	Shoot	Root
Control (C)	0.635 b	0.500 b	22.034 b	8.564 b
C + N (D)[a]	0.717 b	0.470 b	22.269 b	7.429 b
C + *Azolla*	0.802 b	0.515 b	21.332 b	8.874 b
D + *Azolla*	0.562 b	0.420 b	18.556 b	6.694 b
Azolla-soil[b]	1.260 a	0.980 a	37.388 a	16.478 a
LSD (Tukey 5%)	0.345	0.218	9.668	2.906
C.V. %	19.88	17.33	18.19	13.84

Note: Means with different letters, in the same line or column of averages, are significantly different at 5% level of significance by Tukey test.

[a] Ten kilograms of N/ha as NO_3-N.
[b] Preplanting incorporation of *Azolla* (10 kg N/ha).

REFERENCES

1. **Ashton, P. J. and Walmsley, R. D.,** El helecho acuático *Azolla* y su simbionte *Anabaena, Endeavour,* 35 (124), 39, 1976.
2. **Bone, D. H.,** The influence of canavanine, oxygen, and urea on the steady-state levels of nitrogenase in *Anabaena flos-aquae, Arch. Microbiol.,* 86, 13, 1972.
3. **Hardy, R. W. F., Burns, R. C., and Holstein, R. D.,** Application of the acetylene-ethylene assay for measurement of nitrogen fixation, *Soil Biol. Biochem.,* 5, 47, 1973.
4. **Johnson, G. V., Mayeu, P. A., and Evans, H. J.,** A cobalt requirement for symbiotic growth of *Azolla filiculoides* in the absence of combined nitrogen, *Plant Physiol.,* 41, 852, 1966.
5. **Peters, G. A. and Mayne, B. C.,** The *Azolla-Anabaena* relationship. I. Initial characterization of the association, *Plant Physiol.,* 53, 813, 1974a.
6. **Peters, G. A. and Mayne, B. C.,** The *Azolla-Anabaena* relationship. II. Localization of nitrogenase activity as assayed by acetylene reduction, *Plant Physiol.,* 53, 820, 1974b.
7. **Peters, G. A.,** the *Azolla-Anabaena* relationship. III. Studies on metabolic capabilities and a further characterization of the symbiont, *Arch. Microbiol.,* 103, 113, 1975.
8. **Renault, J., Sasson, A., Pearson, H. W., and Stewart, W. D. P.,** in *Nitrogen Fixation by Free-Living Microorganisms,* W. D. P., Ed., Cambridge University Press, London, 1975, 229.
9. **Stewart, W. D. P., Fitzgerald, G. P., and Burris, R. H.,** Acetylene reduction by nitrogen fixing blue-green algae, *Arch. Microbiol.,* 62, 336, 1968.
10. **Subudhi, B. P. R. and Singh, P. K.,** Nutritive value of the water fern *Azolla pinnata* for chicks, *Poult. Sci.,* 57(2), 378, 1978.
11. **Watanabe, I., Berja, N. S., and Alimagno, B. V.,** The Utilization of the *Azolla-Anabaena* Complex as a Nitrogen Fertilizer for Rice, IRRI Research Paper Series No. 11, International Rice Research Institute, Las Baños, Laguna, Philippines, 1977, 1.

Chapter 29

SEASONAL VARIATIONS IN NITROGENASE ACTIVITY OF VARIOUS RICE VARIETIES MEASURED WITH AN *IN SITU* ACETYLENE REDUCTION TECHNIQUE IN THE FIELD

R. M. Boddey and N. Ahmad*

TABLE OF CONTENTS

*   Department of Soil Science, University of the West Indies, St. Augustine, Trinidad.

I. INTRODUCTION

In order to enhance the contribution of nitrogen fixation by bacteria associated with rice roots to the nitrogen economy of the rice crop, several strategies have been put forward. Among them, the selection of rice cultivars which seem to most successfully associate with rhizospheric nitrogen fixers has been suggested as a possibility.

Lee et al.[13] screened 41 rice varieties grown in pots for differences in nitrogenase activity and found considerable and significant differences between many varieties using the acetylene reduction method. Rinaudo[16] similarly found significant differences in acetylene reduction activity (ARA) amongst young seedlings of 28 rice varieties grown in the apparatus of Raimbault et al.[14] Recently Dommergues and Rinaudo[7] have reported that their co-workers studied the ARA of 30 mutants of a single strain of rice variety using this same apparatus. Levels of activity ranged from 861 ± 456 to 7348 ± 1971 n mol C_2H_4/g dry roots per hr. These and similar results do strongly suggest that nitrogen fixation in the rice rhizosphere could be increased by plant breeding. However, as Dommergues and Rinaudo pointed out, the behavior of very young rice plants or rice plants grown in small pots may not reflect the behavior of these same varieties when grown under field conditions.

Watanabe et al.[21] studied the ARA of rice varieties using an *in situ* assay technique[12] at two sites in the Philippines. At one site, although the ARA was measured for two varieties (IR26 and IR36) at several dates during the growth cycle, IR26 was planted 2 months later than IR36. At the other site planting was simultaneous but assays were not, and a statistical comparison could not be applied.

In the work reported here we have compared simultaneously the nitrogenase (ARA) activity of rice varieties assayed several times during the growing season. Experiments were performed at two different sites planted only 1 week apart.

II. MATERIALS AND METHODS

Four varieties of rice were chosen for the field study. Two locally available indica varieties (Joya 47-15-51-9 and Sughandi 47-1-50-9), and two hybrid varieties developed at the International Rice Research Institute (IR5 and IR22) were selected. Field plots were laid out at two sites in Trinidad. One site was situated on a private farmer's land at Munro Road, Charlieville and the other on Cunupia clay at the Ministry of Agriculture Field Experiment Station at El Carmen. Cacandee clay has been described as a Typic Pelludert and Cunupia fine sandy clay as an Aquic Eutropept by Smith.[18] Details of the chemical and physical properties of these soils are available from the authors.

Seedlings were grown at the two sites and transplanted (4 to 8 seedlings per hill) at an age of 5 weeks. All four varieties were planted at the Charlieville site but due to a shortage of the Sughandi seed material only IR5, IR22, and Joya were planted at El Carmen. Transplanting dates were 18th and 25th July 1978 for the Charlieville and El Carmen sites, respectively.

The experiments were layed out in a randomized block design with each variety replicated eight times. Plot dimensions were 3.2 × 2.7 m consisting of nine rows spaced 30 cm apart and hills (subsequently referred to as plants) 20 cm apart within the row. Only every third plant in alternate rows was designated for assay and harvest. At each harvest one plant was chosen at random amongst those designated from each plot. The plots were covered with water throughout the growth period.

In each of the IR22 plots at both sites, a further set of plants was designated for ARA assay but no control of algae with herbicide was attempted. This treatment was regarded as measuring the ARA of both the rhizosphere of IR22 rice and the algae in the surface water.

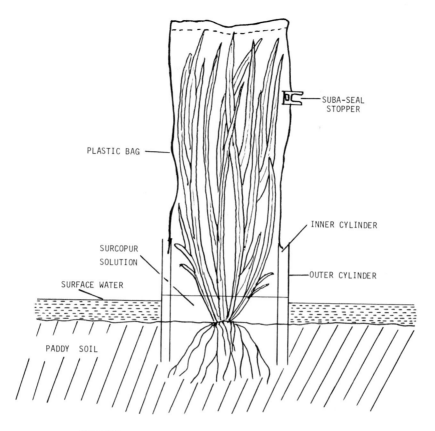

FIGURE 1. Diagram of the apparatus for the *in situ* assay.

A. The *In Situ* Assay

The *in situ* assay technique used in this work was essentially that described in our previous paper[3] with modifications to make it suitable for field work.

The morning before the start of the assay, the plant to be assayed was surrounded by a steel cylinder, 17.5 cm high by 17.5 cm diameter (made from a 1 gal paint tin), and the cylinder pushed a few centimeters into the soil. The surface water within the cylinder was then scooped out and replaced with a 120 ppm solution of the herbicide Surcopur 360 EC to inhibit any algal ARA. The assay chamber consisted of a plastic enclosure made from lie-flat polyethylene tubing, 23 cm wide and 125 μm thick, sealed at one end and the open end stretched over the end of a steel cylinder of 16-cm diameter and 20-cm height made from a 2-kg dried milk tin. A Suba-Seal needle puncture stopper was sealed into the side of the bag as described in our earlier paper.

In the late afternoon the assay chamber was placed over the rice plant by partially deflating the bag and pushing the cylinder a few centimeters into the soil (Figure 1). Acetylene was injected into the bag via the Suba-Seal stopper to a concentration of approximately 20% of the enclosed volume. Either 1.0 or 2.0 mℓ of pure propane was injected as an internal standard.

After an overnight preincubation, an initial sample of gases in the enclosure was taken and stored in a 10-mℓ Vacutainer. After precisely 24 hr a final sample of the gases in the enclosure was taken. The enclosure was then removed and the plant shoot harvested and subsequently dried for 48 hr at 80°C and then weighed.

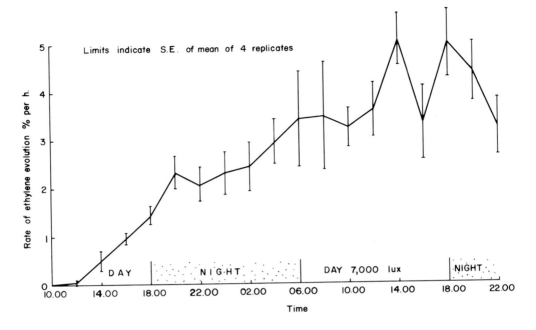

FIGURE 2.   Change in rate of ethylene evolution with time in an assay chamber. Experiment performed in a growth chamber at constant temperature, 27 ± 1°C, 12-hr day (7000 lux), 12-hr night.

The gas samples were analyzed for ethylene, acetylene, and propane, and calculations made for ARA according to Boddey et al.[3]

Two experiments were performed to study the variation in the rate of ethylene production during the assay period. The first was performed in a growth chamber on IR22 rice plants at the tillering stage, grown on Cunupia clay. The growth chamber was maintained at 27 ± 1°C with a 12-hr day (7000 lux) and 12-hr night. The second was performed on IR22 plants in the field at El Carmen 85 days after transplanting. The temperature in the root zone at the base of the stem (4 cm below the soil surface) was simultaneously recorded. In both cases samples of the atmosphere within the enclosure was taken approximately every 2 hr from the time of the injection of acetylene and propane.

III. RESULTS

The variation in the ethylene evolution rate during the *in situ* assay in the constant temperature trial and the field trial are shown in Figures 2 and 3, respectively. The quantity of ethylene produced per hour is expressed as a percentage of the total quantity of ethylene produced during the entire assay period. In this way the variability between plants of different ARA is reduced.

The ARA of each variety at each harvest at the two sites is shown in Figures 4 and 5. Owing to heavy rainfall and lack of suitable drainage, the level of water in the field at Charlieville was often up to 15 cm in depth during the first 4 weeks after transplanting. This accounts for the high variability of the data for this site as some plants barely survived the flooding. It also accounts for the generally lower yields at Charlieville than at El Carmen (Table 1) which is contrary to our experience in previous years.

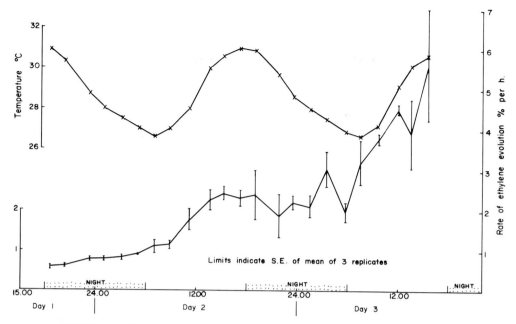

FIGURE 3. Change in rate of ethylene evolution with time in an assay chamber in the field. Temperature measured at the base of the rice stem 4 cm below soil level.

Table 1

INTEGRATED NITROGENASE ACTIVITY AND TOTAL DRY MATTER AND GRAIN YIELDS OF THE VARIOUS RICE VARIETIES AT CHARLIEVILLE AND EL CARMEN

	Charlieville Site				El Carmen Site		
	IR5	IR22	JOYA	SUGHANDI	IR5	IR22	JOYA
''Acetylene Reduced'' during season mmol C_2H_4/m^2	46.8	48.2	37.1	46.1	54.6	54.3	52.0
Total dry matter yield kg/ha[b]	8810	9060	9540	10000	9760	8670	11400
S.E. of mean	±1350	±1350	±1350	±1350	±1190	±1190	±1190
Grain Yield kg/ha[1]	2740	2620	2470	2230	3160	2860	2650
S.E. of mean	±180	±180	±180	±180	±270	±270	±270

By integration of the areas under the curves for each variety (Figures 4 and 5) the total ''acetylene reduced'' per square meter for the season can be estimated based on the figure of 16.7 plants per square meter.

On examination of the data for each harvest it was apparent that in many cases there was a certain degree of correlation between the plant dry matter and the ARA of the plant. To investigate this relationship, regression analysis of the data of each harvest, ignoring varietal effects, was performed. These results are presented in Table 2. To give a further indication of this relationship the mean ARA per gram of dry plant shoot was calculated for each variety at each harvest. The data for the El Carmen experiment are shown in Figure 6.

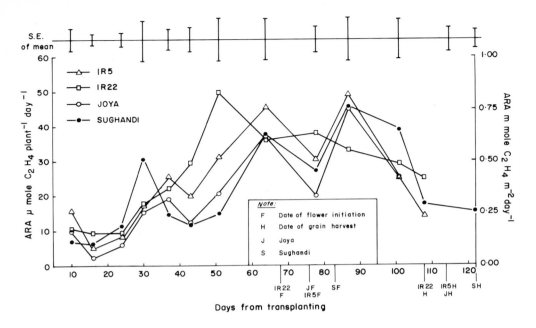

FIGURE 4. Acetylene reduction activity of four rice varieties at the Charlieville site.

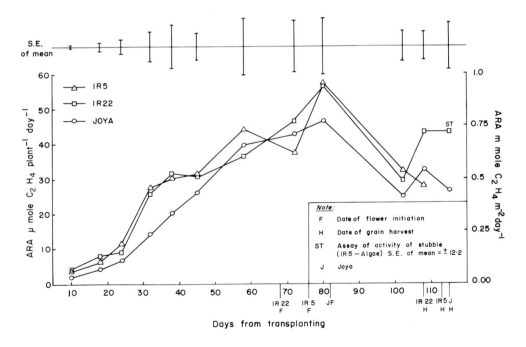

FIGURE 5. Acetylene reduction activity of three rice varieties at the El Carmen site.

In order to determine whether the algal ARA was significantly higher than the IR22 plants where algal activity was inhibited with the herbicide Surcopur, the data was analyzed using the Method of Paired Comparisons as described by Bailey.[1] Only at El Carmen was algal ARA significant for most of the growing season, (Table 3).

Table 2
REGRESSIONS OF ACETYLENE REDUCTION ACTIVITY PER PLANT VS. PLANT DRY WEIGHT

Charlieville experiment[a]

Days from transplanting		10	16	24	30	37	44	51	64	78	86	101	108
Coefficient of determination	R^2%	ns	46.4[b]	24.6[c]	ns	45.3[b]	33.1[b]	68.6[b]	39.8[b]	ns	21.2[c]	23.8[c]	30.9[b]

El Carmen experiment[d]

Days from transplanting		10	18	24	32	38	45	58	72	80	102	108
Coefficient of determination	R^2%	38.7[c]	71.9[b]	53.7[b]	40.3[b]	ns	54.5[b]	23.4[e]	30.6[c]	27.1[c]	33.4[c]	ns

[a] Comparison of 32 pairs of data for each harvest.
[b] Significant at $P = 0.001$.
[c] Significant at $P = 0.01$
[d] Comparison of 24 pairs of data for each harvest.
[e] Significant at $P = 0.05$.

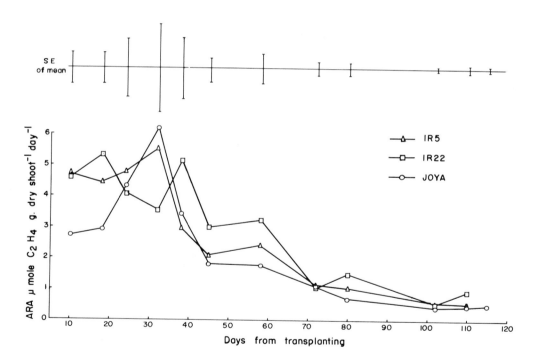

FIGURE 6. Acetylene reduction activity per gram dry shoot of three rice varieties at the El Carmen site.

IV. DISCUSSION

A. The *In Situ* Assay

In the development of an *in situ* acetylene reduction assay technique suitable for use on the lowland rice plant, several difficulties are encountered. Rates of acetylene re-

Table 3
ACETYLENE REDUCTION ACTIVITY OF ALGAE DURING RICE GROWING SEASON[a]

Charlieville experiment

Crop age: days[b]	10	16	24	30	37	44	51	64	78	86	101	108
ARA m mol C$_2$H$_4$/day/m^2	0.33ns	0.37ns	0.99***	1.60*	0.63ns	−0.40ns	−0.55ns	0.12ns	0.15ns	0.16ns	0.41ns	0.02ns
SE of mean	0.28	0.31	0.21	0.61	0.41	0.32	0.41	0.31	0.21	0.23	0.28	0.40

El Carmen experiment

Crop age: days[b]	10	18	24	32	38	45	58	72	80	102	108
ARA m mol C$_2$H$_4$/day/m^2	3.38**	2.17**	1.17**	0.76**	0.51ns	0.85**	1.39*	1.11**	0.11ns	1.01*	1.48ns
SE of mean	0.71	0.69	0.37	0.25	0.38	0.24	0.58	0.35	1.06	0.36	1.06

Note: ns — Not Significant at $P = 0.05$.
  *   Significant at $P = 0.05$.
  **   Significant at $P = 0.01$
  *** Significant at $P = 0.001$.

[a] Data analyzed using the Method of Paired Comparisons (Bailey[1]).
[b] Days from transplanting.

duction in the rhizosphere do not often exceed 2 to 3 μmol ethylene produced per hour and the liberation of small quantities of ethylene into the assay chamber is hindered by absorption of this gas by soil and water.[9,22] For this reason it would be expected that short term assays would produce low and erratic values of ARA in the rhizosphere. The data presented in Figure 2 confirm this and show a rapid increase in ethylene production rate during the first 10 hr of the assay, which we attribute to gradual saturation of the soil and water with evolved ethylene. We suggest that rates measured after this period more nearly reflect the actual nitrogenase activity in the rhizosphere.

Lee and Watanabe[11] suggested that stirring the soil within the chamber would release a significant proportion of the absorbed ethylene. While our experience concurs with this observation, as only one worker was available to set up all of the assay chambers for our experiments, the extra work involved, although not that great, was found to take up too much time. As an alternative, the plants were preincubated overnight (approx. 14 hr) under acetylene to saturate the soil with ethylene before the start of the 24 hr assay.

Long term exposure of nitrogen fixing organisms to acetylene causes an increase in the ARA, due to de-repression of nitrogenase. Long term assay techniques such as that described here have been severely criticized by Hardy et al.[10] and David and Fay[6] for this reason. The data in Figure 3 seem to justify this criticism. After 08.00 hr of day 3 the ARA increased very rapidly, probably due to this effect. Even between 08.00 hr of day 2 and 08.00 hr of day 3 (the 24 hr period used in our assay) the ARA almost trebled.

Short term assays suffer seriously from ethylene absorption losses and in addition do not take into account diurnal fluctuations in activity. Long term assays are much more sensitive and have been found to be more reproducible, but the acetylene reduction rate recorded is almost certainly not a measure of the actual nitrogenase activity at the start of the assay. For this reason the levels recorded here should not be extrapolated to actual levels of nitrogen fixing activity but only be used for strictly comparative purposes.

Superimposed upon the gradual rise of ethylene evolution rate that is recorded in Figure 3, it is possible to discern a marked diurnal fluctuation in this rate. There is some indication that part of this diurnal fluctuation may be due to opening and closing of stomata during light and darkness. This would, respectively, decrease and increase the resistance to the diffusion of ethylene from the rhizosphere into the assay chamber via the plant aerenchyma.[2] This may account for the drop in ethylene evolution rate at the onset of the second dark period in the constant temperature trial. However, it would appear that diurnal fluctuations in temperature are responsible for most of the fluctuation in rate of ethylene evolution in the field. The reasons for this are probably twofold. First, increased temperature increases ARA in the rhizosphere, and second, diffusion of ethylene from the rhizosphere to the chamber atmosphere is faster at higher temperatures.

B. Field Trial Results

In terms of ARA per plant, activity in general was found to be maximal around the time of flowering and grain filling, which is in agreement with the results of other workers.[13,17,21] However, levels of ARA per unit weight of plant shoot were highest in all cases during the first month from transplanting and then steadily declined (Figure 6). Watanabe and Lee[20] expressed their results in a similar manner and, although they found a delay in the onset of ARA (possibly due to high available nitrogen levels in the soil) until the plants were 6-weeks old, levels subsequently declined until harvest time.

Chapter 30

NITROGENASE AND NITRATE REDUCTASE ACTIVITIES IN RICE PLANTS INOCULATED WITH VARIOUS *AZOSPIRILLUM* STRAINS*

F. C. S. Villas Boas and J. Dobereiner**

TABLE OF CONTENTS

*   Translater: Diva Athie.
** Programa Fixação Biologica de Nitrôgênio, Rio de Janeiro, Brasil.

I. INTRODUCTION

Biological N_2-fixation in *Gramineae* in the tropics has been suggested as having great practical potential. Experiments with rice are being developed at International Rice Research Institute (IRRI), Los Baños, Philippines, without the use of fertilizers, and no decrease in soil-N fertility has been noted. On the other hand, the dependence of rice on mineral-N to obtain high yields is well documented. Pereira et al.[11] studied seasonal variation of nitrogenase and nitrate reductase activities in maize (*Zea mays*), while Franco et al.[6] determined these same parameters for beans (*Phasealus vulgaris*). The objective of both studies was to obtain a clearer understanding of how to manipulate the two ways of assimilation to better exploit their potential.

Environmental factors have been mentioned as limiting factors in N_2-fixation.[2] During the growing season of maize and forage grasses, Balandreau[1] noticed that the variation in weekly nitrogenase activity depended mainly on soil mineral-N and humidity, while daily variations could be attributed to differences in soil temperature and light energy. Watanabe and Lee[13] measured a nitrogenase activity peak at flowering, and a decrease during grain filling in irrigated rice. Also with irrigated rice, Balandreau et al.[3] noticed a peak of this enzyme at around 12 to 13 hr while during the night it was depressed.

Baldani and Döbereiner[4] observed specificity in the infection of cereals inoculated with *Azospirillum* spp. The results showed that there was a predominance of *A. brasilense* in the roots of rice sterilized with chloramine for 15 min, suggesting a greater capacity of infection of this group in the interior of the root.

The objective of the present work was (1) to check the behavior of nitrogenase and nitrate reductase activities in the rice plant growth cycle as a function of different amounts of mineral-N; (2) to check the possible effect of inoculation on assimilation; (3) to check the effect of mineral-N on the other parameters described below, and try to find a correlation between them; and (4) to observe the variability in nitrogenase activity in dryland rice, as a function of day-night periods.

II. MATERIAL AND METHODS

Oriza sativa L. was cultivated in the greenhouse in pots of 5 kg, containing hydromorphic type soil from the Ecology Series,[9] at EMBRAPA, Km 47, Rio de Janeiro, Brasil. Six seeds were sown per pot and thinned out to 3 plants per pot, 7 days after germination.

A randomized block design was used with three replications and the following treatments:

1. Three harvest periods: 70, 100, and 120 days after planting
2. Three types of inoculation: control, inoculation with *Azospirillum lipoferum* isolated from maize and inoculation with *A. brasilense* isolated from rice.
3. Four levels of mineral-N as NH_4NO_3: control, 2×15 kg $N_d$ha at planting and 100 days later, and 2×45 kg $N_d$ha at planting and 100 days later.

The plants harvested at 100 days did not receive the second N application, which was added to the plants to be harvested at 120 days. Basic fertilization at planting was given to all pots on the basis of 40 ppm P, 20 ppm K, and minor elements 2 mℓ/pot of the following solution: 15.8 g $CuSO_4 \cdot 5H_2O$; 8.9 g $ZnSO_4 \cdot 7H_2O$; 0.5 g H_3BO_3; 0.5 g $Na_2MoO_4 \cdot 2H_2O$; 20.0 g $FeSO_4 \cdot 7H_2O$; 20 g citric acid; 1000 mℓ H_2O. Soil analysis showed pH 5.1, 0.1 mE Al^{+++}/100 cm^3, 2.9 mE (Ca^{++} + Mg^{++})/100 cm^3 and sandy texture, and no correction was made. Water supply and pest control were given as required.

Two inoculums were used, containing *A. lipoferum* (str[r] 213) and *A. brasilense* (str[r] 219) in concentration of 5.7 × 10^9 cells per milliliter and 6.3 × 10^9 cells per milliliter respectively, both strains coming from the culture collection of EMBRAPA, Km 47. NFb (liquid) + NH_4Cl (10 mM) was used as culture medium for growth of the inoculum, a total of 6 mℓ/plot being applied on the seeds according to the methodology described by Baldani and Döbereiner.[4]

A. Nitrogenase Activity

Evaluation was by acetylene reduction in the intact soil-plant system, using plastic bags, as proposed by Lee and Yoshida.[8] The plants were exposed to gas (10% acetylene, 90% air) for 24 hr. During this period, 0.5 mℓ gas samples were taken 2, 4, 8, 12, 20, and 24 hr after acetylene injection. Production of ethylene was determined by gas chromatograph (Perkin Elmer) equipped with a Porapak N column (3 mm × 2 m) at 110°C and detected by flame ionization.

B. Nitrate Reductase Activity in the Leaves

Neyra and Hageman's[10] method was used. Leaves were taken from three plants of each pot (the last fully developed leaf on each plant) and cut into pieces of approximately 3 × 3 mm. Subsamples of 0.2 g were weighed and placed in glass-flasks containing 5 mℓ of the incubation medium (buffer solution Phosphate (pH 7.5) 0.1 M; KNO_3 0.1 M; Propanol 1% (v/v) Neutronyx (Onyx Co., Jersey City, N.J.) and placed in a dark chamber at 32°C for 1 hr.

The enzyme activity was evaluated by NO_2^- accumulation during the incubation period. Aliquots of 0.2 mℓ were taken from the medium at the beginning (zero time) and at the end of the incubation period (60 min) and 2 mℓ of the solution 1:1 (v/v) 0.02% N-1-naphtyl ethylene diamine and 1% sulphanilamide in 1.5 M HCl was added, the volume being made up to 4 mℓ with distilled water. After 15 min, reading of the absorbance was made at 540 nm in a Coleman colorimeter.

Nitrate reductase determinations were made at 70, 90, and 102 days after planting, which did not coincide with the harvesting periods of the other parameters (70, 100, and 120 days after planting). The plants analyzed 102 days after planting received N-fertilizer as stated, 2 days before the analysis.

C. Dry Matter

The plants were harvested and dried in an oven at 60°C for 72 hr and immediately weighed on an analytical balance.

D. Total N in the Plant

After being dried and weighed, the samples were ground and homogenized. Percent N was analyzed by Kjeldahl method in sub-samples of 0.2 g.

E. NO_3^- Content

The NO_3^- content was determined using the salicylic acid method described by Cataldo et al.[5]

III. RESULTS

A. Nitrogenase Activity

Different levels of mineral-N had a significant effect on nitrogenase activity during

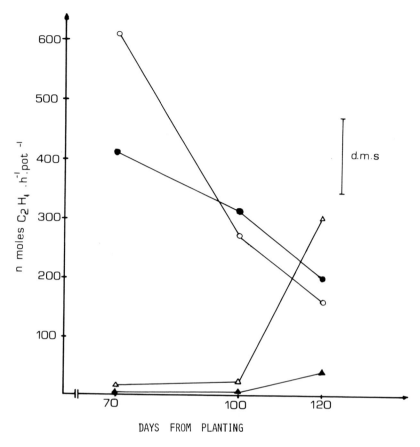

FIGURE 1.   Effect of four levels of mineral-N (0—0 control; ●—● 2 × 15 kg N/ha at and 100 days after planting; ▲—▲ 2 × 30 kg N/ha at and 100 days after planting; Δ—Δ 2 × 45 kg N/ha at and 100 days after planting) on nitrogenase activity at three different periods of the growth cycle. Each value is a mean of nine replications.

the growth cycle of dryland rice. Nitrogenase activity was inhibited in treatments 2 × 30 kg N/ha and 2 × 45 kg N/ha until 100 days after planting; 20 days after the second N application (120 days after planting) there was an unexpected increase in activity. The control and 2 × 15 kg N/ha treatments showed higher activity at 70 days after planting, which coincided with the flowering of the plants, and then declined. The behavior of these two treatments were similar, and contrary to the other two that received higher levels of N (Figure 1). The interaction effect Time × N was highly significant by the F test ($p = 0.05$).

Figure 2 shows nitrogenase variability curves during a period of 24 hr, at two different times (100 and 120 days after planting) and two levels of N (control and 2 × 15 kg N/ha), which demonstrates the depressive effect of night. These data seem to agree with those obtained by Balandreau et al.[3] with irrigated rice.

B. Nitrate Reductase Activity

The effect of inoculation on nitrate reductase activity was significant, but it depended on the time of the growth cycle (interaction time × inoculum). Plants inoculated with *A. brasilense* showed slightly less nitrate reductase activity in the leaves than the control and the plants inoculated with *A. lipoferum*, at the first two determinations

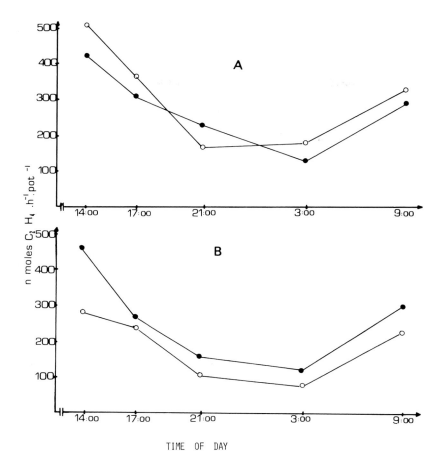

FIGURE 2. Variability in nitrogenase activity over a period of 24 hr, at two levels of mineral-N (0—0 control and ●—● 2 × 15 kg N/ha at and 100 days after planting) and at two harvesting times; (A) 100 and (B) 120 days after planting. Each value is a mean of nine replications.

(70 to 90 days after planting). This situation was reversed at the last analysis (102 days after planting, 2 days after second application of N) (Figure 3). These results are not supported by the literature, and further experiments will have to be made to clarify the mechanism of the activity of these bacteria in the reduction of NO_3 in association with the plant.

Figure 4 shows the variability in nitrate reductase activity as a function of time. Higher levels of nitrate reductase activity corresponded to treatments which received higher amounts of N. At the last determination, 2 days after the second application of N, the nitrate reductase activity increased in all treatments. This activity increased significantly even in the control treatment, which might suggest an increase in rice potential for assimilation through nitrate reductase activity at grain filling.

Different amounts of fertilizer-N have a highly significant effect on the content of NO_3 in the tops and on nitrate reductase and nitrogenase activity, which is shown in Figure 5. With increasing amounts, NO_3 and nitrate reductase increase and nitrogenase activity substantially decreases.

Inoculated treatments did not show significant effects on the analysis of total-N, dry weight and NO_3. In all treatments there was no significant variation in dry weight

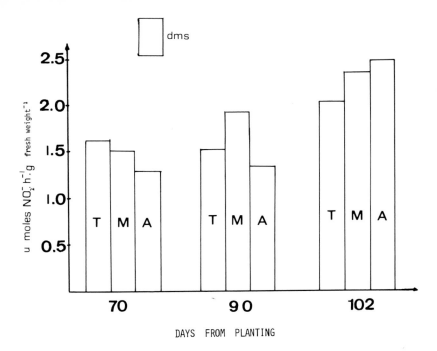

FIGURE 3.   Effect of interaction Time × Inoculation on nitrate reductase activity in the leaves. Three types of inoculation (A) inoculated with *A. brasilense*, (M) inoculated with *A. lipoferum*, and (T) Control. Each value is a mean of 12 replications.

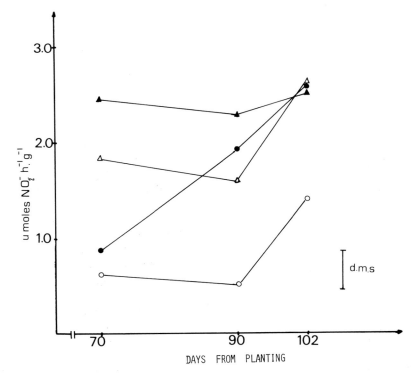

FIGURE 4.   Effect of four levels of mineral-N (●—● control; 0—0 2 × 15 kg N/ha at and 100 days after planting; ▲—▲ 2 × 30 kg N/ha at and 100 days after planting; Δ—Δ 2 × 45 kg N/ha at and 100 days after planting), on nitrogenase activity at three different periods of the growth cycle. Each value in a mean of nine replications.

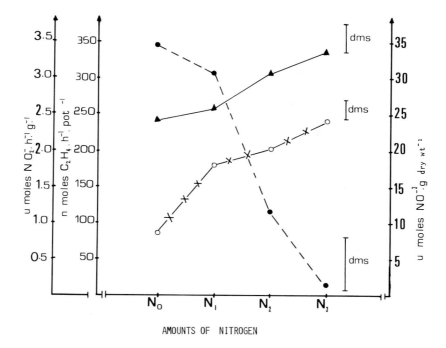

AMOUNTS OF NITROGEN

FIGURE 5.   Variation of NO_3^- content in the aereal part (▲—▲); of nitrogenase activity (●—●—●) and of nitrage reductase activity in the leave (0—×—×—0) as a function of different dosages of mineral-N (NH_4NO_3). N_0 — control, N_1 — 2 × 15 kg N/ha at and 100 days after planting N_2 — 2 × 30 kg N/ha at and 100 days after planting, N_3 — 2 × 45 kg N/ha at and 100 days after planting. Each value is a mean of 27 replications.

after 100 days, while total N increased, reaching maximum values at 120 days after planting (Figure 6 A and B).

IV. DISCUSSION

The plant growth cycle varied between treatments probably due to the different amounts of N, the growth cycle increasing in length with increase of N fertilization. This made the interpretation of results rather difficult because the majority of the parameters and especially the nitrogenase and nitrate reductase activities, changed as a function of the plant growth phase. Pereira et al.[11] and Franco et al.[6] observed in maize and beans, respectively, that nitrogenase activity and nitrate reductase activity potentials have their maximum peaks at different periods of the plant cycle. In the present work, a tendency was noted for nitrate reductase activity to increase at grain filling (102 days after planting) in all treatments (Figure 3).

In sorghum,[12] maize,[11] and rice,[13] maximum nitrogenase activity was found at flowering. In the present experiment, maximum nitrogenase activity values were noted at flowering (70 days after planting) with the lower levels of N (control and 2 × 15 kg N/ha). Nitrogenase activity increased in the last determinations in the treatments that received higher doses of N (Figure 1). This peak is related not only to the growth cycle of the plant, but also to the N-status. Models of seasonal variability of nitrogenase activity and nitrate reductase activity in dryland rice need, however, a larger number of determinations during the growth cycle.

Data obtained for variability in nitrogenase activity over 24 hr, seem to confirm also in dryland rice the depressing effect of night on N_2-fixation.

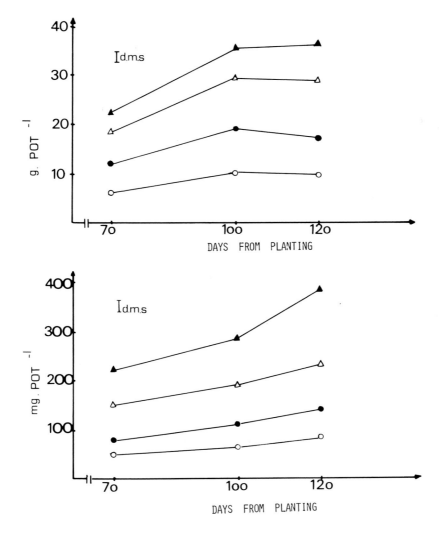

FIGURE 6 (A and B).. Show variation in dry weight and total plant N, respectively, as a function of four levels of mineral-N (0—0) control; ●—● 2 × 15 kg N/ha at and 100 days after planting; Δ—Δ 2 × 30 kg N/ha at and 100 days after planting; ▲—▲ 2 × 45 kg N/ha at and 100 days after planting). Each value is a mean of nine replications.

The behavior of nitrate reductase activity and nitrogenase activity in relation to the variation in plant NO_3^- content, as a function of N-fertilizer application, can be interpreted as due to mechanisms of induction and inhibition, respectively, caused by NO_3^-. Van Berkum and Neyra[12] observed nitrogenase inhibition by NO_3^- in sorghum and maize, while Hageman and Flesher[7] noticed nitrate reductase induction by the same ion.

The effect of strain interaction × time on nitrate reductase activity indicated a specific participation of *Azospirillum* in plant metabolism affecting the reduction and/or translocation of soil NO_3^-, with corresponding response in the nitrate reductase activity in the leaves. A difference in the relationship between plant dry weight and total N as a function of the growing period was clear at the beginning of grain filing (100 days after planting) when dry weight became stable and total N increased (Figure 6 A and

B). This seems to suggest a movement to the N reduction and assimilation systems of the energy hitherto used in the synthesis of dry weight by photosynthesis.

ACKNOWLEDGMENT

The authors thank Mr. Roberto Farias for his collaboration in making the laboratory analyses.

REFERENCES

1. Balandreau, J., Activité nitrogénasique dans la rhizosphère de quelques graminées, Thesis, Université de Nancy, France, 1975.
2. Balandreau, J., Ducerf, P., Fares-Hamad, I., Weinhard, P., Rinaudo, G., Millier, C., Dommergues, Y., Limiting factors in grass nitrogen fixation, *Limitations and Potentials for Biological Nitrogen Fixation in the Tropics,* (Döbereiner, J., Burres, R., and Hollander, A., Eds., Plenum Press, New York, 1978, 275.
3. Balandreau, J., Millier, C. R., and Dommergues, Y. R., Diurnal variations of nitrogenase activity in the field, *Appl. Microbiol.,* 27, 662, 1974.
4. Baldani, V. L. D. and Döbereiner, J., Host plant specificity in the infection of cereals with *Azospirillum* spp., *Soil Biol. Biochem.,* in press.
5. Cataldo, D. A., Haroon, M., Schrader, L. E., and Youngs, V. L., Rapid colorimetric determination of nitrate in plant tissue by nitration of salicylic acid, *Commun. Soil. Sci. Plant Anal.,* 6, 71, 1975.
6. Franco, A. A., Pereira, J. C., and Neyra, C. A., Seasonal patterns of nitrate reductase and nitrogenase activities in *Phaseolus vulgaris* L., *Plant Physiol.,* 63, 421, 1979.
7. Hageman, R. H. and Flesher, D., Nitrate reductase activity in corn seedlings as affected by light and nitrate content of nutrient media, *Plant Physiol.,* 35, 700, 1960.
8. Keuk-Ki, L. and Yoshida, T., An assay technique of measurement of nitrogenase activity in root zone of rice for varietal screening by the acetylene reduction method, *Plant Soil,* 46, 127, 1977.
9. Mendes, W., Lemos, P. de O.e C., Lemos, R. C., Carvalho, L. G. de O., and Rosemburg, R. J., Contribuição ao Mapeamento, em série dos solos do Munícipio de Itaguaí, Bol. nº.12, Inst. Ecol. Exp. Agrícolas, Rio de Janeiro, Brasil, 1954.
10. Neyra, C. A. and Hageman, R. H., Dependence of nitrite reduction on electron transport in chloroplasts, *Plant Physiol.,* 54, 480, 1974.
11. Pereira, P. A. A., von Bülow, J. F. W., and Neyra, C. A., Atividade da nitrogenase, nitro-reductase e acumulação de nitrogênio em milho braqüitico (*Zea mays* L.), *Rev. Bras. Ciência Solo,* 2, 28, 1978.
12. van Berkum, P. and Neyra, C. A., Nitrogenase activity in isolated sorghum roots. Effects of bicarbonate and inorganic nitrogen, *Plant Physiol.,* 57, S-533, 1976.
13. Watanabe, I. and Keuk-Ki, L., Nonsymbiotic Nitrogen Fixation in Rice Paddies, Int. Symp. Biological Nitrogen Fixation in Farming Systems of Humid Tropics, IITA, Ibadan, 1975.

Chapter 31

DISCUSSION ON TERMINOLOGY

Report by: the Editors

The discussion on Terminology, chaired by Dr. Payne, centered on finding a more precisely descriptive alternative to "associative N_2-fixation". It was felt in particular that the use of the term "associative symbiosis" by some authors was quite incorrect. Symbiosis already has a clear meaning: the mutual support of two organisms, and putting the word "associative" as a prefix merely introduces imprecision. Moreover, existing terminology was general, without defining the plant part where the association was situated, yet it was clear that associations in leaves and stems had to be considered as well as with roots.

After some discussion Dr. Nuti suggested "rhizocoenosis", the "coenosis" suffix deriving from the Greek, *koinos,* meaning common in the sense of together. Coenosis was generally agreed and further extensive discussion was concerned with finding suitable prefixes to indicate the nitrogen fixation, and also to define the plant part where the association was located.

The existing word "diazotrophic" was considered the most suitable prefix to indicate nitrogen fixing capacity, and from this was developed "diazotrophic biocoenosis" as a general term to describe nitrogen fixing systems that were associative rather than truly symbiotic in type, associative implying a less well defined form of organization. Then taking the well established prefixes rhizo- (root), phyllo- (leaf) and caulo- (stem), the rest of the terminology followed naturally.

The new descriptive terminology proposed is

1. Diazotrophic biocoenosis, as a general term for associative N_2-fixing biological systems
2. Diazotrophic rhizocoenosis, for associative N_2-fixing systems in, on, or close to the root
3. Diazotrophic phyllocoenosis, for N_2-fixing systems in leaves
4. Diazotrophic caulocoenosis, for N_2-fixing systems in stems

It was resolved that all workers in the field should be asked to adopt the new, more precise, terminology.

Chapter 32

GRASSES, WHEAT, MAIZE, AND SORGHUM — POSITION PAPER

R. V. Klucas and J. Döbereiner

TABLE OF CONTENTS

I. PRESENT SITUATION

Diazotrophic biocoenosis is readily demonstrable in grasses, wheat, maize, and sorghum but the importance of the process to the plant has not been adequately determined. The potential for practical applications of these interactions is significant but basic information is absolutely essential before applied research is advisable.

As Hubbell has pointed out, the demonstration of "significant" N_2-fixation has not always been obtained, but that the word "significant" is relative, and amount of N_2 fixed which may be of no consequence in one situation may be highly significant in another situation. Similarly, successful inoculation experiments have been reported, but inconsistency and failure have been more the rule than the exception. Therefore at the present time inoculation is not a practical agronomic practice.

With grasses, wheat, maize, and sorghum, the diazotrophs are associated either on or inside roots but the site(s) of fixation is still unknown. The causative microbial agents are known in only a few interactions. The better defined ones being *Bacillus* in spring wheat and *Azospirillum* in several plants. *Azospirillum* spp. are commonly found and assumed to be responsible for major diazotrophic and hormonal effects. This microorganism is distributed in soil and plants throughout the world. Recent research has clearly shown that *Azospirillum* exhibits definite plant genotype specificity and possesses a low streptomycin resistance which may be important for infection.

According to DNA-r RNA hybridization experiments by De Smedt, Bauwens, Tytgat and De Ley (personal communication), *Azospirillum* spp. is situated in a r RNA superfamily of bacteria living in close contact with various plant species and not in the group of soil organisms. In this superfamily other organisms such as *Rhizobium, Agrobacterium, Beijerinckia, Rhodopseudomonas, Acetobacter, Gluconobacter, Zymonimas, Xanthobacter,* etc. are grouped. A more extensive study of these genetic similarities might result in a better understanding of diazotrophic biocoenoses vs. symbiotic systems.

Although *Azospirillum* is fairly universally found, the question has been raised as to whether it is always the predominant fixer in situations where its presence is demonstrated. It is necessary to keep in mind that there may be other microorganisms involved as well. The question of nondiazotrophic bacteria in a system has received little attention, and the potential importance of cross feeding between fixers and nonfixers is not established, nor the role that nonfixers might have in modifying the environment for fixers.

Numerous problems still exist in ascertaining the quantitative importance of the process in grasses, wheat, maize, and sorghum. We still must answer such important questions as (1) whether the plant can supply adequate levels of substrates, (2) the mechanism of transfer of substrates and products between plants and diazotraphs, (3) site(s) of nitrogen fixation, (4) the efficiency of the process *in situ,* (5) what root type, "leaking" or "non-leaking" favors diazotrophic rhizocoenosis, there is some evidence in wheat that N_2-fixation is supported best by "nonleaking" roots, and (6) do existing assay methods give realistic appreciation of bacterial populations in certain situations. These are difficult questions to answer but with new information on specificity between plants and microorganisms, meaningful research can be directed possibly to answer these questions.

II. FUTURE WORK

Some major recommendations for the study of diazotrophic biocoenosis in grasses, wheat, maize, and sorghum are

1. Establish criteria for determining the limiting factors in the interactions.
2. Concentrate research efforts on basic aspects to advance our knowledge on physiological considerations of the interaction. This should include the determination of plant factors including root type which favor diazotrophic rhizocoenosis.
3. Identify the microorganisms responsible for nitrogen fixation in each plant system.
4. Establish the efficiency of each system *in situ.*
5. Standardize the methodology for studying diazotrophic biocoenosis. This has to include such methods as isolation techniques, identification procedures and nitrogenase assays. Technical points that need to be considered are: preincubation; the lag factor; the washed root technique; surface sterilization; plate counts from macerated plant tissues, etc.

Chapter 33

RICE — POSITION PAPER

I. Watanabe

TABLE OF CONTENTS

I. PRESENT SITUATION

N_2-fixing microorganisms are present in root, stalks, and leaves of sugar cane, and on the surface of the root and in the rhizosphere soil. Active N_2-fixation does not occur in the above ground parts, as far as is known, but occurs in association with the root, in the rhizosphere soil and in "germinating" pieces of stalk, known as setts, which are used for vegetative propagation. The bacteria present in the stalks therefore provide a means of carrying over the N_2-fixing system from one vegetative generation to another. The relative contribution of "plant" fixation and "soil" fixation is not yet known with any certainty, but accumulating information is tending to indicate "soil" fixation by microorganisms in close relationship with the roots as the main location of nitrogen fixation. Presumably the microorganisms receive their essential energy source from carbohydrates leaked from the roots, possibly together with growth factors. The intensity of microbiological activity in soil being directly related to proximity to the roots is a clear indication of this. Apparently too, compared with either *Phaseolus* or *Zea mays* the sugar-cane root system supports a much higher soil microbiological activity.

The fixation system associated with sugar cane appears to comprise a number of bacteria, with *Azotobacter, Beijerinckia,* and *Bacillus, Clostridia* predominant. *Azospirillum* is present but appears to be of less significance. It is no known whether the system is common to all sugar-cane varieties, or to all areas, nor whether it is affected by soil type or climate. The system may have a certain adventitious element in regard to its composition, dependent on the number and nature of the free-living soil organisms in a particular situation. It is difficult to say which microorganism is responsible for the major fixation, and there is probably a loose association of a number of forms. The possibility of having bacterial subspecies specific to sugar cane has not been ruled out. Evidence for the effectiveness of inoculation is variable though good field results have been reported from India. Work on the bacterial system is somewhat handicapped by the impossibility of obtaining bacteria-free roots or plants except through a long and tedious tissue culture technique, which derives plants from leaf tissue callus.

It is probable that dinitrogen fixation associated with sugar cane has to be considered on a "system" basis, rather than as a one-season phenomenon, as it is a perennial crop, in the sense that ratoon crops are grown for a number of years, and the land is then replanted to cane again. Apart from the fact that seedlings and seed stalks (setts) actively support N_2-fixation we have no knowledge of the distribution of fixing activity during the whole growing period of the cane. Sugar cane in Brazil, retains green leaves until a very late stage (in fact this is a harvesting problem) and it could be that the main N_2-fixation activity in the soil system takes place at a late stage of the plant's growth cycle, when carbohydrate is no longer required for active growth and major accumulation of sugar in the stems takes place, with corresponding root leakage. The nitrogen thus fixed could be visualized as increasing the overall nitrogen balance of the soil system for the following ratoon crop.

II. FUTURE WORK

Quite a lot of information is now available on the associative N_2-fixing system of sugar cane, but our framework of knowledge now requires a great deal of more detailed work, and also increased emphasis on field studies. The information that is particularly required can be summarized as follows:

A. Soil Microbiology

1. Further characterization and identification of the bacteria observed within the root, and other main organisms apparently responsible for associative N_2-fixation needs to be done.
2. What is the relative importance of the "internal" and "external" components of fixation? Are some bacterial specific to sugarcane?
3. Further information on the site of N_2-fixation, the micro-location of fixing organisms, and the relationship to nitrogenase activity is needed.
4. More information on the effect of inoculation with N_2-fixing organisms, such as *Azotobacter* and mixed cultures; do the bacteria move inside the root with root growth, or on the outside?
5. Is inoculation likely to be generally useful, or is there normally a sufficiently wide distribution of appropriate N_2-fixers? What is the situation in new sugar-cane soils, in Amazonas for example?
6. Studies on inoculum production, if inoculation appears a desirable practice, are needed.

B. Genetic Studies

1. We need further genetical evidence and further testing of varieties for capacity for N_2-fixation.

C. Plant Physiology and Biochemistry

1. Clarification is needed of the plant physiological factors associated with N_2-fixation: light, temperature, day length, sugar formation, etc. What is the physio-biochemical factor(s) which determines the capacity or lack of capacity of a plant for N_2-fixation? Should roots be "leaking" or "nonleaking"?
2. What effect does the support of a N_2-fixing system have on the sugar production of a variety? Is N_2-fixing capacity related to varietal differences in sugar content of stems and roots?
3. Clarification is needed of the biochemistry of the N_2-fixation process, especially in relation to potential exudates of carbohydrates and growth factors.
4. Is the movement of gases within the plant a critical factor in N_2-fixation?

D. Field Studies

1. How widespread is N_2-fixation in the field in different sugar-cane growing areas of the world? Is there an effect of latitutde and seasons?
2. What is the potential amount of N_2 fixed under field conditions?
3. What cultural and management practices, fertilization etc., if any, favor or harm N_2-fixation?
4. Further information on the part of the root system that supports N_2-fixation is needed. To what depth is such a system active?
5. In what manner does growth stage, soil moisture, and day and night temperatures affect fixation?
6. Is the lack of response to N-fertilization connected with a high capacity for N_2-fixation, and do those sugar-cane varieties with apparently low fixation have improved response to N-fertilizer?
7. Is one of the reasons for poor N-fertilizer response by sugar-cane in Brazil in part due to the fertilizer having been leached down the profile beyond the rooting zone before the crop has had a chance to fully take it up?
8. Investigations of N_2-fixation in the ratoon crop are required.
9. More experiments are needed on intercropping with sugar cane, e.g., with soybeans, and in some countries the effect of putting sugar cane in a more general cropping system.

Chapter 35

DIAZOTROPHIC BIOCOENOSIS — THE WORKSHOP CONSENSUS PAPER

R. J. Rennie

TABLE OF CONTENTS

I. INTRODUCTION

The concept of dinitrogen-fixing associative symbioses has been subjected to severe criticism in the past. The papers presented at this International Workshop indicate that most of this criticism was unwarranted. The fact that a scientific meeting devoted excluisvely to dinitrogen-fixing associative systems can unite scientists from around the world is perhaps the best testament of how far this field of research has advanced. But perhaps the best prognosis for the future is the fact that international organizations are willing to fund research and support scientific meetings on a subject that still has no immediate agronomic application. This paper represents a consensus of this International Workshop and attempts to define the state of the art and the research priorities for the immediate future.

Microorganisms of the families Spirillaceae (*Azospirillum* spp. and *Campylobacter* spp.), Azotobacteraceae (*Azotobacter* spp., *Derxia* spp., *Beijerinckia* spp.), Enterobacteriaceae (*Enterobacter cloacae, Erwinia herbicola, Klebsiella pneumoniae* spp.) and Bacillaceae (*Bacillus polymyxa, B. macerans, Clostridium* spp.) have been shown to form associations (diazotrophic biocoenosis) with roots (rhizocoenosis,) leaves (phyllocoenosis) or stems (caulocoenosis) of grasses, maize, sugar cane, rice, sorghum, millets, *Spartina* sp., and spring and winter wheat. The macrosymbionts belong to both the C4 and C3 groups of plants with a predominance of the former while the bacteria have little in common except their ability to reduce atmospheric dinitrogen.

II. RESULTS OF THE WORKSHOP

A. Proof of Diazotrophic Rhizocoenosis

In diazotrophic rhizocoenosis, the microorganisms have been shown to invade the plant roots. The exact site of activity of the microorganisms remains unknown since acetylene-reducing activity of surface-sterilized roots is not conclusive proof of the site of dinitrogen-fixing activity. The microorganisms have been shown to reduce both acetylene and dinitrogen in vitro and in the field and this microbially-reduced dinitrogen is, to varying degrees, transferred to the associated plant. Little is known of the actual nature of diazotrophic rhizocoenosis. The initial attraction of the bacteria to the roots may be immunological (like the lectin theory in legumes) or may be physiological and nutritional by virtue of the special preferences of some of the bacteria for five carbon acids. The mode of protection of nitrogenase from oxygen is unknown although, as shown for *Trema* sp., the absence of leghaemoglobin-type compounds does not necessarily preclude dinitrogen fixation. Strict or facultatively-anaerobic dinitrogen-fixing bacteria are the most commonly associated diazotrophs in plants such as wheat which have poor systems of internal gas circulation. Well aerated plants such as maize and sugar cane show a predominance of strict aerobes as the associated dinitrogen-fixer. Once dinitrogen has been reduced microbially, the mechanisms of nitrogen transfer and the timing of this transfer remain unknown. Many of these questions will be answered in the near future presumably by using the *Rhizobium*-legume system as an analogous guide.

Although theoretically, calculations of energetics do not favor the success of diazotrophic rhizocoenosis, significant amounts of dinitrogen fixation associated with plant roots have been shown to occur. The system can fulfill all the basic requirements for dinitrogen fixation: presence of *nif* genes, source of energy and reducing power, requirements for the metallic cofactors Mo and Fe and anaerobiosis.

B. Specificity of Diazotrophic Rhizocoenosis

The dinitrogen-fixing microorganisms demonstrate an affinity for certain plants:

Azotobacter paspali exhibits a specific interaction with *Paspalum notatum; Campylobacter* sp. has been found only in association with *Spartina; Azospirillum brasilense* or *Bacillus polymyxa* are capable of infecting wheat while *Azospirillum lipoferum* has a greater affinity for maize. The best demonstration of host plant specificity is that shown with the reciprocal disomic chromosome substitution lines of spring wheat.

C. Ubiquity of Diazotrophic Rhizocoenosis

Diazotrophic rhizocoenosis has been reported by several investigators in South and North America, Africa, Europe, and Asia (Table 1). The same plants are active in many different ecosystems although their relative activities are often decidely different.

D. Magnitude of Diazotrophic Rhizocoenosis

The acetylene-reducing activities reported in the previous table vary considerably, partly due to different experimental techniques but largely due to the difficulty in obtaining a representative root sample and due to the natural biological variability inherent in the plant. Although these acetylene-reducing activities appear high, when expressed relative to that of soybeans, they are relatively insignificant. This is not surprising since the calculated field efficiencies of dinitrogen fixation (Table 2) are very low in nonleguminous systems other than blue-green algae. The participants in the Uppsala symposium attempted, using their best available field data, to obtain an estimate of the true significance of diazotrophic rhizocoenosis and concluded, that although high variability existed, 10 to 20 kg N-fixed/ha/year was a reasonable figure. However, in proper perspective, even this small amount of dinitrogen-fixing activity represents a significant input of nitrogen into nitrogen-deficient ecosystems that have no other source of exogenous nitrogen. Thus, although the fixation rates associated with grasslands are low relative to those of legumes (Table 3), by virtue of the large land areas in nonleguminous cultivation, diazotrophic rhizocoenosis represents a highly significant contribution to the total world nitrogen balance. Data presented at this Workshop support the premise that diazotrophic rhizocoenosis results in the reduction of at least 30 kg N/ha/year.

It is essential that we should know the amount of field fixation in order to be in a better position to consider the possibilities and economics of plant breeding for better N_2-fixation, bearing in mind the long term nature and cost of plant breeding. The ultimate objective is not completely clear. Do we aim for N-fertilizer replacement or aim for replacing only part of N-fertilizer? Both may be suitable targets for different situations, the former in low-intensity farming where cost of fertilizer is a major factor.

III. FUTURE RESEARCH PRIORITIES

It has been documented that agronomically significant amounts of dinitrogen fixation do occur in association with certain plants and not with others. The ability of a plant to support associated dinitrogen-fixing bacteria can be altered by genetic manipulation of the plant genome in the case of spring wheat.

The increased use of ^{15}N combined with proper taxonomic identification of the associated dinitrogen-fixing bacteria has resulted in a greater understanding of the nature and the magnitude of diazotrophic rhizocoenosis. The major contribution of this meeting has not been the introduction of startling new knowledge, but rather the presentation of concise, well-documented evidence for the existence and the agronomic significance of diazotrophic rhizocoenosis.

A great deal of knowledge is still required for diazotrophic rhizocoenosis to be a valuable tool for practical agriculture. Techniques of identifying and quantifying pres-

Table 1
SOME COMPARISON OF SPECIES SUPPORTING ASSOCIATIVE DINITROGEN FIXATION (ACETYLENE REDUCTION)

Country	Species	N_2ase Activity (nmol C_2H_4/g root/hr)
Brazil	*Brachiaria mutica*	150—750
	B. rugulosa	5—150
	Cymbopogon citratus	10—100
	Cynodon dactylon	20—270
	Cyperus rotundus	10—30
	Digitaria decumbens	20—400
	Hyparrhenia rufa	20—30
	Melinis minutiflora	15—40
	Panicum maximum	20—300
	Paspalum notatum	2—300
	Pennisetum purpureum	5—1000
	Saccharum spp.	5—50
	Sorghum vulgare seedlings	10—100
Ivory Coast	*Andropogon* spp.	50—380
	Brachiaria brachylopha	100—140
	Cyperus obtusiflorus	30—620
	C. zollingeri	50—160
	Cyperus sp.	1150—1900
	Panicum maximum	100—530
	Loudetia simplex	0—32
	Pennisetum purpurem	8
	Paspalum sp.	430
	Eleusine indica	374—850
	Bulbostylis sp.	1750
Nigeria	*Andropogon gayanus*	15—270
	Cenchrus ciliaris	0—16
	Cymbopogan giganteus	60—85
	Cynodon dactylon	10—50
	Cyperus sp.	2
	Hyparrhenia rufa	30—140
	Panicum maximum	75
	Pennisetum coloartum	60
	Pennisetum typhoides	13
	Sorghum vulgare	3—195
		22—83
North America	*Triticum aestivum*	220
	Avena sativa	10
	Agrostis tennuis	130
	Bromus inermis	10
	Festuca bromoides	30
	Digitaria sanguinalis	20—580
	Panicum virgatum	270
	Sporobolus heterolepsis	810
Senegal	*Oryza sativa*	1190—2360
	Eleusine coracana	2—380
	Paspalum virgatum	1032

Note: Some discretion should be used in comparing these figures, as some systems were intact, others nonintact, and samples refer to both incubated and nonincubated systems.

Table 2
CALCULATED FIELD EFFICIENCIES OF N₂-FIXATION

kg N/ha

Legumes

Lucerne	300
Clover	150
Lupin	150
Pulse	55

Nonlegumes

Alder	100
Blue-green algae	25
Azotobacter spp.	0.3
Clostridium spp.	0.3

Table 3
N-FIXATION IN THE BIOSPHERE

Land use	Fixation rate (kg N fixed/ha/year)	Amount fixed (ton/year ($\times 10^6$))
Legumes	140	35
Rice	30	4
Other crops	5	5
Grassland	15	45
Forest	10	40
Unused	2	10
Total land		139
Sea	1	36
Total		175

From Paul, E. A., *Ecol. Bull. Stockholm*, 26, 282, 1978. With permission.

ent and new diazotrophic rhizocoenoses must be improved. The acetylene-reduction technique, although analytically sensitive and inexpensive, is a short-term kinetic assay. The difficulties of root sampling combined with diurnal and seasonal variations in acetylene-reducing activities plus the imprecision of the arbitrary 3:1 nitrogen: acetylene ratio, make extrapolation to field conditions highly questionable.

^{15}N₂ reduction techniques are the only absolute proof of dinitrogen fixation and its increased used has been essential to the progress of this field of science. Still, it is also a short term kinetic assay, which is 1000 times less sensitive than the acetylene-reduction technique and its only real use should be as an absolute proof of dinitrogen fixation. ^{15}N isotope dilution techniques yield a true integrated value for dinitrogen fixation. The use of this technique in the field must be encouraged since it is not only an absolute proof of dinitrogen fixation but also the only truly quantitative means of estimating dinitrogen fixation under field conditions. The greatest difficulty in employing this technique will be the ability to have a true nonfixing plant as a control. For this purpose the recognition of genotypes which positively do not support diazotrophic biocoenosis would be valuable.

$\delta^{15}N^{0}/_{00}$ data permit the estimation of dinitrogen fixation at levels of ^{15}N natural abundance. This has several advantages since the ecosystem need not be disturbed by the addition of labeled fertilizer nitrogen and the great cost of purchasing ^{15}N is avoided. The great natural variation in $\delta^{15}N^{0}/_{00}$ limits the sensitivity of the procedure but data presented at this Workshop indicate a true potential of the $\delta^{15}N^{0}/_{00}$ procedure for scanning for new diazotrophic biocoenosis under field conditions. A combination of these four techniques in the field will enable the absolute proof and quantification of diazotrophic rhizocoenosis.

Enzymatic interferences that limit the magnitude or efficiency of dinitrogen fixation must be controlled. The inefficiency of H_2 evolution by nitrogenase must be harnessed to increase the efficiency of the nitrogenase reaction. Every effort should be made to use only nar$^-$ nir$^-$ dinitrogen fixing bacteria to eliminate denitrification.

Massive commercial field inoculation studies are premature. However, scientists must ascertain the effect of inoculation with dinitrogen-fixing microorganisms on dry matter and nitrogen yield of the associated plants. The nitrogen content (%N) is not a meaningful estimate of inoculation effect since it is subject to dilution due to yield. Inoculated microorganisms should be genetically marked with at least two markers to facilitate their reisolation and to maintain strain purity.

Although tremendous emphasis has been placed on *Azospirillum* spp. in diazotrophic rhizocoenosis, evidence was presented at this Workshop for the importance of *Azotobacter* spp. with sugar cane, *Bacillus* spp. with wheat, *Campylobacter* sp. with *Spartina* and Enterobacteraceae with rice. Increased emphasis should be placed on both numbers and dinitrogen-fixing activities of all indigenous diazotrophic microorganisms found associated with plants. Particular emphasis should be placed on the possibility of additional microorganisms, either diazotrophic or not, being involved in diazotrophic rhizocoenosis in a supporting role.

IV. CONCLUSIONS

Data presented at this meeting has convincingly substantiated earlier claims of the agronomic significance of diazotrophic biocoenosis. The need for increasingly rapid communication between the scientists working in this rapidly developing field resulted in the call for a newsletter perhaps in conjunction with the *Rhizobium Newsletter*. Further, the scientists agreed that biannual meetings should be held devoted exclusively to diazotrophic biocoenosis.

AUTHOR INDEX

SUBJECT INDEX